John Lubbock

The Scenery of Switzerland and the Causes to which it is Due

Vol. I

John Lubbock

The Scenery of Switzerland and the Causes to which it is Due
Vol. I

ISBN/EAN: 9783744740487

Printed in Europe, USA, Canada, Australia, Japan

Cover: Foto ©Andreas Hilbeck / pixelio.de

More available books at **www.hansebooks.com**

COLLECTION
OF
BRITISH AUTHORS
TAUCHNITZ EDITION.

VOL. 3211.

THE SCENERY OF SWITZERLAND.
BY
THE RIGHT HON. SIR JOHN LUBBOCK, BART., M.P.

IN TWO VOLUMES.—VOL. I.

THE SCENERY

OF

SWITZERLAND

AND THE

CAUSES TO WHICH IT IS DUE

BY

THE RIGHT HON.
SIR JOHN LUBBOCK, BART., M.P.
F.R.S., D.C.L., LL.D.

COPYRIGHT EDITION.

IN TWO VOLUMES. VOL. I.

WITH ILLUSTRATIONS.

LEIPZIG
BERNHARD TAUCHNITZ
1897.

PREFACE.

In the summer of 1861 I had the pleasure of spending a short holiday in Switzerland with Huxley and Tyndall. Tyndall and I ascended the Galenstock, and started with Benen, who afterwards lost his life on the Haut de Cry, up the Jungfrau, but were stopped by an accident to one of our porters, who fell into a deep crevasse, from which we had some difficulty in extricating him, as Tyndall has graphically described in his *Hours of Exercise on the Alps.*

From that day to this many of my holidays have been spent in the Alps. On them I have enjoyed many and many delightful days; to them I owe much health and happiness, nor must I omit to express my gratitude to the Swiss people for their kindness and courtesy.

My attention was from the first directed to the interesting problems presented by the physical geography of the country. I longed to know what forces had raised the mountains, had hollowed out the lakes, and directed the rivers. During all my holidays these questions have occupied my thoughts, and I have read much of what has been written about them. Our knowledge is indeed very incomplete, many problems still baffle the greatest Geographers, as to others there is still much difference of opinion. Nevertheless an immense fund of information has been gathered together; on many points there is a fair concensus of opinion amongst those best qualified to judge, and even where great authorities differ a short statement of their views, in a form which might be useful to those travelling in Switzerland, could hardly fail to be interesting and instructive. No such book is, however, in existence. I urged Tyndall and several others far better qualified than I am myself, to give us such a volume, feeling sure that it would be welcome to our countrymen, and add both to the pleasure and to the interest of their Swiss trips. They were all, however, otherwise occupied, but they encouraged me to attempt it, promising me their valuable assistance, and this must be my excuse for undertaking the task, perhaps prematurely. Tyndall we have unfortunately lost, but Professor Heim and Sir John Evans have been kind enough to take the trouble of looking through the proofs, and I am indebted to them for many valuable suggestions.

The Swiss Government have published a series of excellent maps, which has been prepared at the cost of the State, under the general direction of General Dufour. There is also a geological map by Heim and an older one by Studer and Escher, which was admirable at the time it appeared, and has in the main stood the test of more recent researches. Studer was in fact the father of Swiss Geology; he accumulated an immense number of observations which have been most useful to subsequent authors, and if I have not quoted his researches more often, it is because I have been anxious to give the latest authorities. In 1858 he suggested that the Dufour map should be taken as the basis of a geological survey on a larger scale. To this the Swiss Government assented; they voted the modest sum of £120, since increased to £400 a year, and appointed a Commission, consisting of Messrs. B. Studer, P. Mirian, A. Escher von der Linth, A. Favre, and E. Desor. Under their supervision the present geological map, in twenty-five sheets, has gradually appeared: the last being published in 1888, on the very day of Studer's death.

In addition to the geological maps themselves, the Commission have published a splendid series of descriptive volumes, over thirty in number, by A. Müller, Jaccard, Greppin, Möesch, Kaufmann, Escher, Theobald, Gilliéron, Baltzer, Fritsch, Du Pasquier, Burckhardt, Quereau, Heim, Schmidt, Favre, Renevier, Gerlach, Schardt, Fellenberg, Rolle, Taramelli, and others.

This is not the place to catalogue the separate

Volumes and Memoirs on Swiss Geology and Physical Geography. Jaccard in his work on the Jura and Central Switzerland enumerates no less than 959, but among the most important I may mention Heim's magnificent work, *Mechanismus der Gebirgsbildung,* Studer's *Geologie der Schweiz,* Agassiz's *Études sur les Glaciers,* Suess's *Das Antlitz der Erde,* Favre's *Recherches Geologiques;* for the fossils that of Heer; and among shorter publications, in addition to those by the geologists already referred to, particularly those of Bonney, Morlot, Penck, Ramsay, Rütimeyer, and Tyndall.

I have dwelt specially on the valleys of the Arve, Rhone and Rhine, the Reuss, Aar, Limmat and Ticino as types of longitudinal and transverse valleys; and because they are among the districts most frequently visited. They have, moreover, been admirably described, especially by Favre, Heim, Renevier and Rütimeyer.

I am fully conscious of the imperfections of this book: no doubt by waiting longer it might have been made better; but I should have felt the same then also, and in the words of Favre, "il n'y a que ceux qui ne font rien qui ne se trompent pas."*

* *Rech. Geol.* III. 76.

CONTENTS
OF VOLUME 1.

CHAPTER I.
THE GEOLOGY OF SWITZERLAND.   Page

The influence of geology on scenery.—Difficulty of subject.—Geological history of Switzerland.—Igneous rocks.—Gneiss.—Origin of Gneiss.—Granite.—Porphyry.—Protogine.—Serpentine. Crystalline Schists.—Carboniferous period.—Ancient mountain chain.—Permian period.—Triassic.—Jurassic; Lias, Dogger, Malm.—Cretaceous.—Tertiary; Eocene, Flysch, Nummulitic beds; Miocene.—Summary 25

CHAPTER II.
THE ORIGIN OF MOUNTAINS.

Continents the true mountain ranges.—Two classes of mountain ranges: table mountains and folded mountains.—Peaks, two classes of: volcanoes, mountains of denudation.—Origin of mountain ranges, cooling and consequent contraction of the earth.—Crust consequently either broken up or folded.—Hence table mountains and folded mountains.—Table mountains.

—Cape of Good Hope.—Horsts.—Alps due not to
upheaval, but to folds.—Amount of compression.—
The Jura.—Amount of denudation.—Dip and strike.
—Faults.—Anticlinals and synclinals.—Folding of solid
rock, proof of.—Fractured fossils.—Cleavage . . . 49

CHAPTER III.
THE MOUNTAINS OF SWITZERLAND.

General direction of pressure.—The range due to folding,
the separate summits being parts which have suffered
least from denudation.—The Rhone-Rhine Valley.—
Geotectonic valleys, and valleys of erosion.—Transverse ranges.—Enormous amount of denudation.—
The Secondary strata formerly extended over the summits.—12,000 feet of strata probably removed from
Mont Blanc.—General section of Switzerland.—Folds,
inversion, and overthrusts.—Earthquakes.—The Plain
of Lombardy an area of sinking 81

CHAPTER IV.
SNOW AND ICE.

Snow-fields.—The snow-line.—Firn or Névé.—Red snow.
—Depth of snow.—Beauty of snow-fields.—Avalanches.
—Glaciers.—Structure of glacier ice.—Glacier grains.
—Movement of glaciers.—Rate of movement.—Cause
of movement. — Regelation. Crevasses. — Veined
structure. — Dirtbands. — Moulins. — Moraines. — Ice
tables.—The glacier of the Rhone.—Beauty of glaciers 100

CHAPTER V.
THE FORMER EXTENSION OF GLACIERS.

Evidence of the former extension of glaciers.—Moraines
and fluvio-glacial deposits.—Ancient moraines, distri-

bution of.—Erratic blocks.—Polished and striated surfaces.—Scratched pebbles.—Upper limit of ancient glaciers.—Flora and fauna.—Evidence of milder interglacial periods.—Limits of the ancient glaciers.—Probable temperature of the Ice age.—Table of glacial deposits 130

CHAPTER VI.
VALLEYS.

Valleys not all due to rivers.—Geotectonic valleys.—Valleys of subsidence.—Plain of Lombardy.—Valley of the Rhine near Basle.—Classes of valleys.—Longitudinal valleys.—Synclinal valleys.—Anticlinal valleys.—Combes.—Transverse valleys.—The river system of Switzerland.—Two main directions.—Cirques.—Weather terraces.—Age of the Swiss valleys . . . 167

CHAPTER VII.
ACTION OF RIVERS.

Three stages in river action: deepening and widening; widening and levelling; deposition.—River gorges.—River cones.—Cones in the Valais.—Cone of the Borgne.—Cones and villages.—Slope of a river.—River terraces.—Valley of the Ticino.—The Rhine.—Val Camadra.—Effects of floods.—Giants' caldrons . 188

CHAPTER VIII.
DIRECTIONS OF RIVERS.

Lake district.—Plateau of Lannemazan.—Main directions of Swiss rivers.—The rivers of the Swiss Lowlands.—Rivers and mountains.—The Rhone and its tributaries.—Changes in river courses.—The tributaries of the

	Page
Rhine.—Former course of the Rhone.—The Danube.—Age of the Swiss rivers	212

CHAPTER IX.
LAKES.

Height and depth of Swiss lakes.—Classes of lakes.—Lakes of embankment, of excavation, and of subsidence.—Crater lakes.—Corrie lakes.—Lakes due to rockfalls.—The great lakes.—Theories of Ramsay, Tyndall, and Gastaldi.—Lakes in synclinal valleys.—Lakes due to moraines.—Lakes due to changes of level of the land.—The Italian lakes.—Colour of the Swiss lakes.—The Beine or Blancford 234

CHAPTER X.
THE INFLUENCE OF THE STRATA UPON SCENERY.

Character of scenery dependent on weathering, the climate, the character and inclination of the rocks.—Siliceous rocks.—Calcareous rocks.—Argillaceous rocks.—Gneiss.—Granite.—Crystalline Schists.—Porphyry.—Dolomite.—Karrenfelder.—Glaciated scenery.—Moraine scenery.—Rockfalls.—Earth pyramids . . . 258

TABLE OF LINEAR MEASURE.

1 inch = 25.40 millim.
1 foot = 0.305 metre.
1 yard = 0.914 metre.
1 mile = 1.609 kilom.

LIST OF ILLUSTRATIONS.

Fig.		Page
1.	Cascade of Arpenaz	56
2.	Diagram in illustration of folded mountains. (After Ball)	59
3.	Hall's experiment illustrating compression. (After Geikie)	60
4.	Diagram showing the artificial folds produced in a series of layers of clay on indiarubber. (After Favre)	60
5.	Section across the Jura from Brenets to Neuchâtel. (After Jaccard)	63
6.	Section from Basle across the Alps to Senago, north-west of Milan. (After Rütimeyer)	64
7.	Section of the Tremettaz. (After Favre and Schardt)	65
8.	Diagram showing the "strike" on the ground-plan A and the "dip" in the section B. (After Prestwich)	66
9.	Monoclinal fold	67
10.	A fault. (After Geikie)	67
11.	Fold-fault. Line of fault at the upper displaced bed. (After Heim and De Margerie)	68
12.	An inclined fold. (After Heim and De Margerie) .	68
13.	Razed folds— a, anticlinal; b, synclinal	69
14.	Diagram showing anticlinal and synclinal folds . .	69

Fig.		Page
15.	Hand specimen of contorted mica schist. (After Geikie)	74
16.	Section of röthidolomite. (After Heim)	76
17.	Piece of stretched verrucano	76
18.	Stretched and broken belemnites, half size. (After Heim)	77
19.	A fragment of nummulitic limestone. (After Heim)	78
20.	Section of compressed argillaceous rock in which cleavage structure has been developed. (After Geikie)	80
21.	Section of a similar rock which has not undergone this modification. (After Geikie)	80
22.	Carboniferous trough on the Biferten Grat (Tödi). (After Rothpletz)	82
23.	Profile through the gneiss masses between the Rhone at Viesch and the Averserthal. (After Schmidt)	91
24.	Section from the Spitzen across the Ruchen to the Maderanerthal. (After Heim)	92
25.	Section across the Mont Blanc range. (After Favre)	94
26.	Section across the Alps. (After Heim)	95
27.	Diagram showing motion of a glacier. (After Tyndall)	114
28.	Section of icefall and glacier below it, showing origin of veined structure. (After Tyndall)	116
29.	Diagram showing the flow of glacier ice. (After Tyndall)	117
30.	Sketch map of the Mer de Glace. (After Tyndall)	118
31.	View of the Grimsel	131
32.	Scratched boulder	132
33.	Diagram showing moraine and fluvio-glacial strata. (From *Le Syst. Glaciaire des Alps*)	134
34.	Figure representing river terraces and glacial deposits in the valley of the Aar a short distance above Coblenz. (From *Le Syst. Glaciaire des Alps*)	135

LIST OF ILLUSTRATIONS.

Fig.		Page
35.	Map of the country between Aarau and Lucerne	136
36.	Diagram showing crag and tail. (After Prestwich)	148
37.	View of the Brunberghörner and the Juchlistock near the Grimsel, showing the upper limit of glacial action. (After Baltzer)	149
38.	Section of combe. (After Noë and De Margerie)	169
39.	Do. do.	169
40.	Do. do.	170
41.	Section from the valley of the Orbe to Mont Tendre. (After Jaccard)	171
42.	Sketch map of the Swiss rivers	175
43.	Diagram in illustration of mountain structure	181
44.	Diagram illustrating weather terraces in the valley of the Bienne (Jura). (After Noë and De Margerie)	184
45.	Do. do. do.	185
46.	Diagram showing the course of a river through hard and soft strata	185
47.	The normal slope of a river	188
48.	Diagrammatic section of a valley	192
49.	River Cone. Front view	196
50.	Do. Side view	197
51.	Map showing junction of Rhone and Borgne	199
52.	Profiles of the principal rivers in the valley of the Garonne. (After Noë and De Margerie)	200
53.	Slope of the principal rivers in the valley of the Garonne. (After Noë and De Margerie)	201
54.	Section across the valley of the Ticino. (After Bodmer)	202
55.	Diagram showing river terraces in Val Camadra. (After Heim)	203
56.	Diagram of a river valley. Section representing harder calcareous rock overlying a softer bed. (After Noë and De Margerie)	204

FIG.		Page
57.	Diagram to illustrate a river now running on an anticlinal	216
58.	Sketch map of the Rhone and its tributaries	219
59.	River system round Chur, as it is	220
60.	River system round Chur, as it used to be	221
61.	Section showing river terraces in the Oberhalbsteinrhein. (After Bodmer)	223
62.	Section across the Val d'Entremont. (After Bodmer)	229
63.	Section across the Val d'Entremont from Six Blanc to Catogne. (After Bodmer)	230
64, 65, 66.	Diagrams to illustrate Corrie Lakes	239
67.	Diagram to illustrate the action of rivers and glaciers	244
68.	Diagram section along the Lake of Geneva. (After Ramsay)	245
69.	Diagram illustrating the side of a lake. (After Forel)	256
70.	Diagram showing the needleforms of the granite ridge of the Gauli. (After Baltzer)	266

LIST OF STRATA.

		PRINCIPAL SWISS REPRESENTATIVES.
Recent		
Post-Tertiary		Glacial and Interglacial deposits
Tertiary	Pliocene	
	Miocene	Mollasse and Nagelfluc
	Eocene	Nummulitic Limestone and Flysch
Secondary	Cretaceous	Cenomanian (Seewenkalk)
		Gault
		Schrattenkalk, Urgonian, and Aptian
		Neocomian
		Valangian
	Jurassic	Malm (Hochgebirgskalk)
		Dogger
		Lias
	Trias	Keuper
		Muschelkalk. Haupt Dolomite
		Bunter Sandstein
Palæozoic	Permian	Verrucano
	Carboniferous	Puddingstones, Slates, and Sandstone
	Devonian?	
	Silurian?	Various Crystalline Schists
	Cambrian?	Eruptive Rocks
Crystalline Schists Gneiss, etc.		

2*

GLOSSARY.

Anchitherium. An Eocene quadruped, intermediate between the Tapirs and the Equidæ. They are regarded as ancestors of the Horse.

Anticlinal, see p. 69.

Archæan. The Geological Record may be classified in 5 divisions—1, Archæan; 2, Palæozoic (Ancient Life); 3, Secondary or Mesozoic (Middle Life); 4, Tertiary; and 5, Quaternary.

Argillaceous Rock. Consisting of, or containing, clay.

Basalt. A black, extremely compact igneous rock, which breaks with a splintery or conchoidal fracture.

Batrachia. The group of animals to which Frogs, Toads, and Newts belong.

Belemnites. Cephalopods; allied to the Squid and Cuttlefish.

Bergschrund, see p. 102.

Bündnerschiefer, see p. 37.

Bunter, see p. 34.

Carboniferous, see p. 32.

Cargneule. A rock belonging to the Triassic period.

Cleavage, see p. 78.

Crevasses, see p. 114.

Deckenschotter, see p. 162.

Dinotherium. A gigantic Mammal belonging to the Miocene period.

Diorite. A rock differing from granite in containing less Silica.

Dip, see p. 66.

Dogger, see p. 37.

Dolomite. Magnesian Limestone.

Eocene, see p. 41.

Erratics, see p. 43.

Eyed-Gneiss, see p. 28.

Felspar. Constitutes the largest portion of Plutonic and Volcanic rocks; anhydrous, aluminous, and magnesian Silicates.

Firn, see p. 101.

Flysch, see p. 42.

Fold Fault, see p. 68.

Foraminifera. A group of microscopic shells.

Gabbro, see p. 28. A group of coarsely crystalline rocks.

Gault, see p. 40.

Geotectonic, see p. 167.

Glacier-Grain, see p. 108.

Gneiss, see p. 26.

Granite, see p. 29.

Hauptdolomite, see p. 35.

Hochgebirgskalk, see p. 38.

Hornblende. A group of Silicates, so called from their horn-like cleavage, and peculiar lustre.

Horst, see p. 54.

Keuper, see p. 34.

Lias, see p. 36.

Læss, see vol. II. p. 41.

Magma, see p. 28.

Malm, see p. 38.

Mastodon. A gigantic quadruped, allied to the Elephant.

GLOSSARY.

Mesozoic, see under *Archæan*.
Miocene, see p. 43.
Mollasse, see p. 43.
Monoclinal Fold, see p. 66.
Moraine, see p. 121.
Muschelkalk, see p. 35.
Nagelflue, see p. 43.
Neocomian, see p. 39.
Névé, see p. 101.
Nummulites, see p. 41.
Orthoclase. A form of Felspar. An original constituent of many crystalline rocks, including Granite, Gneiss, Syenite, etc.
Outcrop, see p. 66.
Palæotherium. A Tapir-like Mammal, belonging to the Eocene period.
Palæozoic, see under *Archæan*.
Permian, see p. 34.
Plagioclase. A kind of Felspar. Tschermak characterises it as a mixture of Soda Felspar and Lime Felspar.
Plutonic. Igneous rocks which have consolidated below the surface.
Porphyry, see p. 29.
Protogine, see p. 29.
Quartz. A form of Silica.
Regelation, see p. 112.
Schist. A rock which is split up into thin irregular plates.
Secondary, see under *Archæan*.
Seewen-Limestone, see p. 40.
Sericite. A talc-like variety of Mica.
Serpentine, see p. 30.
Shale. A rock which splits along laminæ of deposition.
Slate. A rock which splits along lines of cleavage.
Slickensides, see p. 75.
Strike, see p. 66.

Syenite. A mineral composed of Felspar and Hornblende.
Synclinal, see p. 69.
Trachite. A lava with a low percentage of Silica.
Trias, see p. 34.
Urgonian, see p. 40.
Valangian, see p. 39.
Verrucano, see p. 33.

CHAPTER I.

THE GEOLOGY OF SWITZERLAND.

> Vidi ego, quod fuerat quondam solidissima tellus,
> Esse fretum: vidi factas ex æquore terras;
> Et procul a pelago conchæ jacuere marinæ.
> <div style="text-align:right">Ovid, <i>Metam.</i> xv. 262.</div>

> Straits have I seen that cover now
> What erst was solid earth; have trodden land
> Where once was sea; and gathered inland far
> Dry ocean shells.
> <div style="text-align:right"><i>Ovid's Metam.</i>, trans. by H. King.</div>

The Scenery of Switzerland is so greatly due to geological causes, that it is impossible to discuss the present configuration of the surface, without some reference to its history in bygone times. I do not, however, propose to deal with geology further than is necessary for my present purpose.

The subject presents very great difficulties, not only because the higher regions are so much covered with snow, accessible only for a few weeks in the year, and in many places covered by accumulations of debris, but especially because the rocks have been subjected to such extremes of heat and pressure that not only have the fossils been altered, and often entirely destroyed, but the very rocks themselves have been bent, folded, reversed, fractured, crushed, ground, and so completely metamorphosed that in many cases their whole character has been changed beyond recognition.

Igneous Rocks.—Gneiss.

To commence with the Igneous series, which come from the fiery heart of the earth. Gneiss, which is in Switzerland as elsewhere the fundamental rock, forms in great part the central ranges, reappearing also here and there in other parts, as for instance on the Rhine at Laufen, and would, it is thought, be found everywhere if we could penetrate deep enough.

Gneiss is composed of Quartz, Felspar, and Mica, with a more or less foliated structure. The Felspar is generally white, but sometimes green or pink, and has often a waxy lustre; the Mica is white, brown, or black. The Quartz forms a sort of paste wrapping round the other ingredients.

Gneiss presents the same general characters all over the world. It is not all of the same age, and if some is comparatively recent, at anyrate the oldest rock we know is Gneiss. This gives it a peculiar interest. The foliation of Gneiss is probably of two kinds: the one due to pressure, crushing, and shearing of an original igneous rock such as Granite, the other to original segregation-structure.*

"Gneiss," says Bonney, "may be, if not actually part of the primitive crust of the earth, masses extruded at a time when molten rock could be reached everywhere near to the surface."** When the crust of the earth first began to solidify the waters of the present ocean must have floated in the atmosphere as steam, so that even at the surface there would be a pressure equal to more than 12,000 feet of water. The cooling also must have been very slow. Still, the original crust, if we use the words in their popular sense to mean the superficial layers, was probably more like basalt, or the lavas of our existing volcanoes. Gneiss, on the other hand, must have cooled and solidified under considerable pressure and at a great depth. When we stand on a bare surface of Gneiss we must remember—and it is interesting to recollect—that it must have been originally

* Heim, *Beitr. z. Geol. K. d. Schw.*, L. XXIV.; Geikie, *Text-book of Geology*.
** *Story of our Planet*.

covered by several thousand feet of rock, all of which have been removed.

"Probably," says Geikie, "the great majority of geologists now adopt in some form the opinion, that the oldest or so-called 'Archæan' Gneisses are essentially eruptive rocks. . . . Whether they were portions of an original molten 'magma' protruded from beneath the crust or were produced by a refusion of already solidified parts of that crust or of ancient *sedimentary* accumulations laid down upon it, must be matter of speculation." *

On the other hand, Gneiss is certainly not all of the same age, since in some instances it traverses other strata. There appear, moreover, to be cases in which *sedimentary* strata have been metamorphosed by heat or pressure into a rock which cannot mineralogically be distinguished from Gneiss.

Gneiss presents many varieties. The principal are Granite-gneiss, where the schistose arrangement is so coarse as to be unrecognisable, save in a large mass of the rock; Diorite-gneiss; Gabbro-gneiss, composed of the materials of a Dolerite or Gabbro, but with a coarsely schistose structure; Porphyritic-gneiss or Eyed-gneiss, in which large eye-like kernels of Orthoclase or Quartz are dispersed through a finer

* *Text-book of Geology.*

matrix, and represent larger crystals or crystalline aggregates which have been partially broken down and dragged along by shearing movements in the rock.

GRANITE.

Granite, like Gneiss, is composed of Quartz, Mica, and Felspar, but differs from it in not being foliated.

Granite is a plutonic rock and may be of any age; it often sends veins into the surrounding strata, which it then forces out of position, in which case they show evidence as they approach it of being much altered by heat. It solidified at a considerable depth below the surface, and its upper portions probably flowed out as lava. It presents much variation: if it shows traces of foliation it is known as Gneiss-granite. Hornblende-granite contains Hornblende in addition to the other elements. Diorite differs in containing Plagioclase instead of Orthoclase, and less Silica; if the Felspar crystals are large and well defined, it is known in popular language as Porphyry. Syenite consists of Felspar (Orthoclase), Hornblende, and a little Quartz. Protogine, so named because it was formerly supposed to be the oldest of all rocks, is a Granite, containing Sericite instead of, or with, ordinary Mica.

Granite, like Gneiss, must have solidified under considerable pressure, and therefore at a great depth. In the first place, the crystals it contains could not have been formed unless the process of cooling had been very slow. In addition to this, they present a great number of minute cavities containing water, liquefied carbonic acid, and other volatile substances. Sorby, whose main conclusions have since been verified by others, has endeavoured to calculate what must have been the pressure under which Granite solidified, by measuring the amount of contraction in the liquids which have been there imprisoned. He considered that the Granites which he examined must have consolidated under pressure equivalent to that of from 30,000 to 80,000 feet of rock. The more superficial layers probably resembled Basalt.

Serpentine.

Serpentine is a compact or finely granular rock, olive-green, brown, yellow, or red, and has a more or less silky lustre. There has been much doubt as to its origin, but it is now regarded generally as an altered igneous rock.

Crystalline Schists.

Over the Gneiss lie immense masses of Crystalline Schists, several thousand feet in thickness.

No fossils have been found in them, though the presence of Graphite and seams of Limestone have been supposed to indicate the existence of vegetable and animal life. The more ancient were perhaps deposited while the waters of the ocean were still at a high temperature. So generally distributed are these Schists, that in the opinion of many geologists they everywhere underlie the other stratified formations as a general platform or foundation. In parts of Switzerland, however, sedimentary strata have been so much modified by pressure, and in many cases by heat, that it is very difficult, and even in places impossible, to distinguish them from the older Crystalline Schists. "At one end," says Geikie, "stand rocks which are unmistakably of sedimentary origin, for their original bedding can often be distinctly seen, and they also contain organic remains similar to those found in ordinary unaltered sedimentary strata. At the other end come coarsely crystalline masses, which in many respects resemble Granite, and the original character of which is not obvious. An apparently unbroken gradation can be traced between these extremes, and the whole series has been termed Metamorphic from the changed form in which its members are believed now to appear." The discovery of fossils has indeed proved that certain Schists are Silurian, others Devonian, Carboniferous, and even Jurassic, but no Swiss geologists consider that the Crystalline Schists of the Central "Massives" of the Alps are metamorphic Mesozoic rocks.* The Schists are generally intensely folded and crumpled. The presence of boulders of foliated Crystalline Schist in the Carboniferous Puddingstones, proves that the foliation was original, or at least anterior to the Coal period.**

The problems, however, presented by these rocks are, as Geikie says, so many and difficult that comparatively little progress has yet been made in their solution.

The Carboniferous Period.

The earliest fossiliferous rocks in Switzerland belong to the Coal or Carboniferous period. The older Cambrian and Silurian rocks, which elsewhere present so rich a flora and fauna, and attain a thickness of many thousand feet, are perhaps represented in Switzerland by some of the Crystalline Schists, though this is not yet certainly proved.

A belt of Carboniferous strata extends from Dauphiné along the valley of the Isère and the Arve,

* Heim, *Quart. Jour. Geol. Soc.* 1890.
** Lory, *Int. Geol. Cong.* 1888.

presenting fossiliferous deposits at Brevent, Hüningen, etc. It then passes along the lower Valais, and, if the Verrucano belongs to this period, occupies a considerable part of the district between the Upper Rhine and the Walensee. It is clear, however, and this indeed applies to the fossiliferous strata generally, that these beds are only remnants of much more extensive deposits. In places they have been removed, and in others they have been deeply buried under more recent strata. At the same time much of Switzerland is supposed to have been land at this period, probably forming a large island, or islands, while the presence in the Valais and the Mont Blanc district of Puddingstone containing pebbles and boulders shows that there must have been some high land, and rapid streams. The Coal was probably formed in deposits somewhat similar to our peat-mosses.

The vegetation consisted principally of Ferns, Mosses, Clubmosses (Lycopodiacæ), and Equisetums. There appear to have been some flowering plants, but the blossoms were probably inconspicuous. Insects were represented by forms resembling the Cockroach, but there were no Bees, Flies, Butterflies, or Moths. Batrachia make their appearance, but there were no Mammals or Birds. The Verrucano, or, as it is often called, Sernifite, from the Sernfthal, is a sandy or pebbly deposit belonging either to the close of the Carboniferous or commencement of the Permian period.

Permian.

During the Permian period also Switzerland was partly above the sea-level, partly covered by the sea. The land appears to have gradually sunk, commencing in the east, and in the

Triassic

period the sea appears to have covered the whole area of Switzerland. The name "Trias" was given to it because in many districts, though not everywhere, it falls into three principal divisions, a brown, white, green, or reddish Sandstone, known as the Bunter Sandstein, the Muschelkalk or Shelly Limestone, and the Keuper, consisting of marls and limestones.

In Switzerland, as in England, there are considerable salt deposits belonging to this period. Another very characteristic rock of this age is Gypsum, and the Dolomites also belong to this period. Many mineral waters spring up from, and owe their properties to, the Triassic beds. The Keuper districts are generally rich, Dolomites on the contrary poor, desolate, and often almost without vegetation, but very beautiful from their richness of colour, and rugged forms.

The Muschelkalk is often, as, for instance, on the Virgloria pass, a hard black limestone, splitting into thin slabs, which take a good polish and are used for tables.

The earliest Mammals appeared in this period.

To the Upper Trias belongs a thick deposit of grey, whitish, or yellow Dolomite, sometimes compressed into a grey or blackish Marble, which is known as Hauptdolomite, and, especially to the east of the Rhine, from its great durability often forms the highest and wildest ridges of the mountains. It is unfossiliferous.

The account here given of the geography of Switzerland in past times differs, as will be seen, considerably from that indicated in the maps to Heer's *Primæval World of Switzerland*. Prof. Heer regarded the present boundaries of the different formations as indicating their original extension. This however is certainly not the case. The Jurassic strata, for instance, were not deposited near any land. There are no shore animals nor pebbles, as there must have been if they were coast deposits.

JURASSIC.

The principal Jurassic strata in Switzerland are the Lias, the Dogger, and the Malm. They attain together a thickness of over 2500 feet. During this period Ammonites and Belemnites reached their fullest development, as also did the great Sea-reptiles, the Icthyosaurus and Plesiosaurus. At this period also flourished the flying reptiles or Pterodactyles, and we also meet the first bird (Archæopteryx), which differed from all existing species, by the possession of a long tail, and in other ways.

Lias.

During this period the whole of Switzerland appears to have been covered by the sea. There must however have been land not very far off, as remains of Beetles, Cockroaches, Grasshoppers, Termites, Dragon-flies, Bugs, and other Insects occur in the Lias of Schambelen, near the junction of the Reuss, the Aar, and the Limmat, and elsewhere. No Bees, Butterflies, or Moths have been met with.

It is probable that the Black Forest and the Vosges were dry land. The fossils, however, on the whole, indicate a deep sea. The Lias is grey or blackish, calcareous, sandy, or argillaceous stratum. The dark color is probably owing to the amount of organic matter which it contains. Heer suggests that the best explanation may be afforded by the *Sargasso Sea*. The Atlantic Ocean, for an area of about 40,000 square miles, is covered by Sargasso-weed so densely that ships sometimes find a difficulty in forcing their way through it. The sea is deep, and the fragments of dead weed are probably quite decayed before they reach the bottom, to which they would give a dark colour. He thus explains the colour of this limestone.

The "Bündner Schiefer" so largely developed in the Grisons and Valais are now considered, from the fossils which have been discovered in several places, to belong to this period. It it probable however that the strata marked as Bündner Schiefer on the Swiss maps do not all belong to the same period.

Dogger or Brown Jura.

Switzerland was for the most part under water at this period, but that there must have been land in the neighbourhood during some part of the time is proved by the existence, near Porrentruy, of beds containing several species of Limpets (Patella), Periwinkles (Purpura), Mussels (Mytilus), Neritas, and other shore molluscs. It is probable that the Black Forest and the Vosges were then dry land.

Malm, or Upper Jurassic.

The Malm is characterised by a considerable development of Coral reefs, which often attained a great thickness. Between the Corals, which in some cases still retain their natural position, are many remains of Sea-urchins, Sponges, Molluscs, and some Crustacea, united by calcareous cement into a more or less solid rock. They are often beautifully preserved, having been embedded in the soft mud of a quiet sea, which extended completely over the Central Alps. Indeed the southern shore of the Jurassic Sea must, in Heim's opinion, be looked for in northern Africa.

The Malm is yellow and white in the Jura, blue-black in the Alps; by its hard, bare, steeply inclined rocks, and dry sterile slopes, it gives a special character to the landscape, while the Dogger, and still more the Lias, from their numerous marly layers, furnish a very fertile soil. Where Malm is a dark-bluish, grey, conchoidal, calcareous rock, it is known as "Hochgebirgskalk." In the celebrated deposits of Solenhofen many remains of *Insects* occur, including a Moth, the earliest Lepidopterous insect yet known.

CRETACEOUS.

As in the Jurassic period, so also in the Cretaceous, Switzerland was under the sea. To the east, however, was dry land. The complete difference between the animals of the Malm or Upper Jurassic, and then of the Neocomian or Lower Cretaceous, appear to imply a change of conditions or great lapse of time. It was at one time supposed* that the southern shore of the Swiss Cretaceous Sea followed a line drawn from the Walensee to Altorf, the Lake of Brienz and Bex, but though this is the present limit of the strata they once extended much farther, and have been removed by denudation. Heim considers that islands began to show themselves in the region of the Central Alps in Cretaceous times.

The Swiss Cretaceous strata fall into five principal divisions. The first or oldest—Valangian—consists of a dark hard silicious and sometimes oolitic limestone as on the Sentis, or of bluish grey marls and limestone as in the Jura.

The Neocomian, from the old name of Neuchâtel, is sometimes a dark gray or black hard marl, sometimes a bluish grey marl which easily disinte-

* Heer, *Primæval World of Switzerland*.

grates in the air, but contains beds of excellent stone of which Neuchâtel is built.

The Urgonian (so called after the town of Orgon, near Arles), or Schrattenkalk, is widely distributed in the Alps. It is a hard white limestone, the surface of which is often furrowed by innumerable channels, which form a perfect labyrinth. It stands in rocky walls often several hundred feet high, and from its great powers of resistance often forms the ridges and water-sheds. It is arid and barren, offering a great contrast to the Neocomian, which generally bears a luxuriant vegetation.

The Gault contains many dark green grains which are a silicate of protoxide of Iron. It forms the dark bands which are so conspicuous against the paler colour of the other Cretaceous rocks.

The Seewen Limestone, so called from the village of Seewen on the Lake of Lowerz, corresponds in age to our Chalk, and like it consists mainly of microscopic shells. The eastern and western parts of Switzerland differ considerably in the species. The Cretaceous deposits being of marine origin we cannot expect to know much of the land animals or plants. The forests, however, contained Cycads and Conifers, Pines, Sequoias, etc., and Dicotyledonous trees now make their appearance, the earliest being a species resembling a poplar found in the Cretaceous beds of Greenland. In the upper Cretaceous strata Dicotyledons are more numerous, and it is interesting to find that they are mostly species in which the pollen is carried from flower to flower by the wind, or such as Magnolia, which are fertilised by beetles. Bees and Butterflies were still apparently absent or rare, and hence also the beautiful flowers specially adapted to them.

Eocene.

At this period the formation of islands on the site of the present Alps appears to have commenced. The two principal rocks of the Eocene period are the Nummulitic Limestone and the Flysch. They represent differences of condition rather than of time. Bands of Nummulitic Limestone often occur in the Flysch, showing that for a while the sea was favourable for the development of Nummulites. Then the conditions changed, and they disappeared. This happened again and again.

Nummulitic Limestone.

The Nummulitic Limestone is so called because it contains numerous Foraminifera, the shells of which are in some species so flattened that they resemble pieces of money. In many cases, moreover, the size increases the resemblance. The sea in which they lived was of great extent. The pyramids are built of Nummulitic Limestone, and the Nummulites are traditionally said to be the petrified remains of the lentils on which the children of Israel were fed by Pharaoh. They occur also in Asia Minor, Persia, on the Himalayas, and in Thibet, where they now rise to a height of 5000 metres.

Flysch.

The Flysch is a very remarkable and important deposit. The name is a local Bernese expression, which was adopted by Studer. Flysch is sometimes marly, sometimes calcareous, sometimes sandy. It is often slaty, and is extensively worked. It attains a thickness of nearly 2000 metres, and is evidently marine, but except in the slates of Matt, the only fossils found in it have been certain impressions which have been supposed to be Seaweeds, or perhaps Worm burrows. What are the conditions under which these have been preserved when all other traces of organic remains have perished, is a mystery. The Flysch mountains present soft outlines, and their slopes support a rich carpet of vegetation.

These are the two principal deposits of the

Eocene period, so far as Switzerland is concerned. In other strata numerous fossils have been found, including many Mammalia, and even a Monkey.

Miocene.

During this period the main elevation of the Alps took place. We should naturally expect that rapid rivers would rush down from the heights bringing masses of gravel with them, and in fact we find enormous deposits of coarse gravel, often cemented into a hard rock, and containing blocks six inches, a foot, and even sometimes as much as a yard in diameter. This conglomerate is known as the Nagelflue, and the materials of which it is composed become gradually finer as we recede from the Alps, forming a more or less marly deposit known as the Mollasse. It attains a great thickness; indeed the whole of the Rigi from the Lake of Lucerne to the summit consists of Nagelflue. The Mollasse is composed of several deposits, some fresh-water and some marine; it is probable that the conditions may have been different in different parts of what are now the Swiss lowlands. The pleasant scenery of Central Switzerland is greatly due to the Mollasse. The Freshwater Mollasse is generally soft, but the Marine beds afford excellent building materials. Large quantities are brought to Zürich from the upper part of the lake. It contains beds of brown coal and is rich in fossils. Indeed the deposits at Oeningen contain perhaps the richest collection of fossils in the world. Taking the Miocene period as a whole we know nearly 1000 species of plants and 1000 insects; of reptiles 32 species have been discovered, whereas in Switzerland now there are only 27. As regards Mammals 59 have been determined, while at present Switzerland contains 62; but though the numbers are so nearly the same, the species are all different and belong to very different groups. Of the present species 15 are bats, but no bat has been found in the Swiss Miocene. It contains on the other hand no less than 25 Pachyderms. The Wild Boar is the only present representative of the order, but during the Miocene period Tapirs and tapir-like Palæotheria, the horse-like Anchitherium, two species of Mastodon, the Dinotherium and no less than 5 species of Rhinoceros, roamed over the Swiss woods and plains. Of plants we know already 1000 species. Many resemble, and are probably ancestral forms of, those now flourishing in very distant parts of the world. Thus there are several Sequoias, one of which (Sequoia Langsdorfii) closely resembles the Redwood of California, and another (Sequoia Sternbergii) the gigantic Wellingtonia. Another species resembles the Marsh Cypress of the southern United States. There are also Australian types such as Hakeas and Grevilleas, while Palms, Liquidambars, Cinnamon, Figs, Camphor trees, and many other southern forms also occur. Of Oaks Prof. Heer has described no less than 35 species.

Moreover many of the Miocene plants have been found in the far North, implying a comparatively uniform and mild climate. Thus Sequoia Sternbergii is abundant in the lignites of Iceland, and Sequoia Nordenskioldi has been found in Greenland. As a whole the Flora resembles that of the present day, but represented by types now scattered over the whole world, and has most affinity with that of North America, as it contains over 200 North American against 140 European types. They have as a rule small and wind-fertilised flowers. Those which are more conspicuous, and which add so much beauty to our modern flora, were less numerous in Miocene times; and many families are altogether absent, such as Rosaceæ, Crucifers, Caryophyllaceæ, Labiatæ, Primulaceæ, etc. Bees and Butterflies, though already existing, had not yet so profoundly modified and developed the flowers. The Miocene species were all killed off or driven south by the Glacial

Period and succeeded by others better able to stand a cold climate. There was on the contrary no such complete change in the marine flora and fauna.

Summary.

Looking at the Alps as a whole the principal axis follows a curved line from the Maritime Alps towards the north-east by Mont Blanc, Monte Rosa,* and St. Gotthard to the mountains overlooking the Engadine.

The geological strata follow the same direction. North of a line running through Chambery, Yverdun, Neuchâtel, Soleure, and Olten to Waldshut on the Rhine are Jurassic strata; between that line and a second nearly parallel and running through Annecy, Vevey, Lucerne, Wesen, Appenzell, and Bregenz on the Lake of Constance, are the lowlands, occupied by later Tertiary strata; between this second line and another passing through Albertville, St. Maurice, Leuk, Meiringen, and Altdorf lies a more or less broken band of older Tertiary strata, south of which are first a Cretaceous zone, then one of Jurassic age, followed by a band of crystalline rocks, while the central core, so to say, of the Alps, consists mainly of Gneiss or

* This name has no reference to colour, but is derived from "reuse," a local name for glacier.

Granite. If we draw a line across Switzerland, say from Basle to Como, we find from Basle to Olten, say to the line of the Aar, Jurassic formations thrown into comparatively gentle undulations, and stretching from south-west to north-east. From Olten to Lucerne, the great plain of Switzerland is made up of upper Tertiary strata, known as Mollasse, and Nagelflue, consisting of sand and gravel washed down from the rising mountains and deposited partly in a shallow sea, partly in lakes. Near Lucerne we come upon Eocene strata, also of marine origin, which have been raised to a height of as much as 2000 metres.

Continuing in the same direction, and soon after passing Vitznau, we come upon Cretaceous rocks, which occupy most of the canton of Nid Dem Wald. In Ob Dem Wald we find ourselves on Jurassic. In other parts of Switzerland a considerable thickness of Triassic strata appears beneath the Jurassic, and rests on Verrucano, one of the Carboniferous series, but along our line the Jurassic region is immediately followed by Crystalline rocks, and Gneiss, forming the great Central ridge of Switzerland, and reaching as far as the Lake of Como. On the south of the mountain range, as on the north, the Gneiss is followed in succession by Carboniferous, Triassic, Jurassic, Cretaceous, and Tertiary strata, but they form narrower belts. Bellagio is on Trias; from the Island of Comacino the Gulf of Como is surrounded by Jurassic strata, south of which is a band of Cretaceous, running from the Lago Maggiore, opposite Pallanza, by Mendrisio, Como, Bergamo, and the south end of the Lago d'Iseo to Brescia, and so on further to the east.

Speaking roughly then we may say that the backbone of Switzerland consists of Gneiss and Granite, followed on both sides by Carboniferous, Triassic, Jurassic, Cretaceous, and Tertiary strata. These however are all thrown into a succession of gigantic folds, giving rise to the utmost complexity. The similarity of succession on the two sides of the ridge gives reason for the belief that the Triassic, Jurassic, and Cretaceous strata north and south of the Alps were once continuous, and this impression is confirmed by other evidence, as will be shown in the following chapters.

CHAPTER II.

THE ORIGIN OF MOUNTAINS.

> There rolls the deep where grew the tree.
>   O earth, what changes hast thou seen!
>   There, where the long street roars, hath been
> The stillness of the central sea.
>
> The hills are shadows, and they flow
>   From form to form, and nothing stands;
>   They melt like mist, the solid lands,
> Like clouds they shape themselves and go.
>   TENNYSON.

THE true mountain ranges, that is to say, the elevated portions of the Earth's surface, are the continents themselves, on which most mountain chains are mere wrinkles; nevertheless when we speak of mountains, we mean as a rule those parts of the land which stand high relatively to the sea-level.

Mountain ranges in this sense may be classed under two main heads,* viz.:—

---

* I say "main" heads, because in certain cases there may be other explanations. Von Richthofen has suggested that the Dolomites of the Tyrol were originally coral reefs.

Scenery of Switzerland. I.

I. Table mountains.
II. Folded mountains.

The highest points or peaks may be again divided into two classes—volcanoes, and those due to weathering.

Volcanoes.

Volcanoes have had comparatively little effect on the scenery of Switzerland. There is only one group of hills in Switzerland, those of Hohgau near the Lake of Constance, which is of Volcanic origin.

There are indeed certain isolated masses of igneous rock, as for instance in the Chablais, and again near Lauchern in Wandelibach, which are probably the necks of ancient Volcanoes.

Mountains of Denudation.

Let us imagine a country raised above the water with a gradual and uniform slope towards the sea. Rivers would soon establish themselves, guided by any inequalities of the surface, and running at more or less equal intervals down to the water level. They would form valleys, down the sides of which secondary rivulets would flow into the main streams. The rain and frost would denude with especial rapidity those parts of the surface which offered the least effective resistance, and thus not only would the original watershed be cut into detached summits, but secondary ridges would be formed approximately at right angles, to be again cut into detached summits like the first.

The general opinion of geologists used, however, to be, in the words of Sir R. Murchison, that "most of the numerous deep openings and depressions which exist in all lofty mountains were primarily due to cracks which took place during the various movements which each chain has undergone at various periods."

In support of this view such gorges as those of Pfäffers, the Trient, the Gorner, the Aar, etc., were quoted as conclusive cases, but even these are now proved to have been gradually cut down by running water.

The rapidity of denudation is of course affected greatly by the character of the strata, so that the present level depends partly on the original configuration, partly on the relative destructibility of the rock. The existing summits are not those which were originally raised the highest, but those which have suffered the least. And hence it is that so many of the peaks stand at about the same level. Everyone who has ever stood at the top of such a mountain as the Piz Languard, which I name as being so easily accessible and so often visited, must have been struck by this fact; and must have noticed that the valleys are a far less important part of the whole district than they seem when we are below. The Matterhorn is obviously a remnant of an ancient ridge, which gives the peculiar straight line at the summit. The noble mass of the Bietschorn again, which forms such an imposing object as we look down the valley of St. Niklaus across the Rhone at Visp, is a part of the surrounding granite which has resisted attack more successfully than the rest of the rock. The mountain crests, solid as they look from a distance, are often covered by detached fragments, shattered by storms, and especially by frost.

Mountain Ranges.

The present temperature of the Earth's surface is due to the Sun, that supplied from the original heat of the planet being practically imperceptible. The variations of temperature due to seasons, etc., do not extend to a greater depth than about 10 metres. Beyond that we find as we descend into the Earth that the heat increases on an average about 1° Fahr. for every 50 metres.* Even, there-

* Agassiz, however, in the case of the Calumet Mine near Lake Superior, found a rate of 1° Fahr. for every 223 ft. (*Amer. Journ. of Science*, 1895).

fore, at comparatively moderate depths the heat must be very great. Many geologists in consequence, have been, and are, of opinion that the main mass of the Earth consists of molten matter. We know, however, that the temperature at which fusion takes place is raised by pressure, and it must not, of course, be assumed that the temperature continues to increase so rapidly beyond a certain depth. Other great authorities,* therefore, are of opinion that the mass of the Earth, though intensely hot, is solid, with, no doubt, lakes of molten matter. In either case the central mass continues slowly to cool and consequently to contract. The crust, however, remains at the same temperature and consequently of the same dimensions. This being so, under the overwhelming force of gravity one of the two things must happen. Either (1), parts of the crust must break off and sink below the rest; or (2), the surface must throw itself into folds.

Table Mountains.

Where the first alternative has happened we find more or less numerous faults.

Those parts which have not sunk, or which have

* See, for instance, Lord Kelvin, *Lectures and Addresses*, vol. II.

sunk less than the rest, remain as tabular mountain masses, more or less carved into secondary hills and valleys by the action of rain and rivers. Such, for instance, is the Table Mountain of the Cape of Good Hope; its relative height is not due to upheaval, but to the surrounding districts having sunk.

As the crust of the Earth cooled and solidified, certain portions "set," so to say, sooner than others; these form buttresses, as it were, against which the surrounding areas have been pressed by later movements. Such areas have been named by Suess "Horsts," a term which it may be useful to adopt, as we have no English equivalent. In some cases where compressed rocks have encountered the resistance of such a "Horst," as in the northwest of Scotland and in Switzerland, they have been thrown into most extraordinary folds, and even thrust over one another for several miles.

Murchison long ago expressed his surprise at the existence of great plains such as those of Russia and Siberia. L. v. Buch suggested as a possible explanation that they rested on solid masses which had cooled down early in the history of the planet, and thus had offered a successful resistance to the folds and fractures of later ages.

Folded Mountains.

The Swiss mountains, however, belong to another class, and have a very different character. They are greatly folded and compressed (see Figs. 23-26). Fig. 1 represents the Cascade of Arpenaz in the valley of the Arve. It shows a grand arch, but does not include the whole fold, which takes the form of an **S**, the middle part only being included in the photograph.

It used to be supposed that mountains were upheaved by forces acting more or less vertically from below upwards, and the igneous rocks which occupy the centre of mountain ranges were confidently appealed to in support of this view. It must be confessed that when we first visit a mountainous region, this theory seems rational and indeed almost self-evident. It is now, however, generally admitted that such an explanation is untenable; that the igneous rocks were passive and not active; that, so far from having been the moving force which elevated the mountains, they have themselves been elevated, and that this took place long after their formation. Near the summit of the Windgälle, in the Reuss district, is, for instance (Fig. 24), a mass of Porphyry. The eruption of this Porphyry must have taken place be-

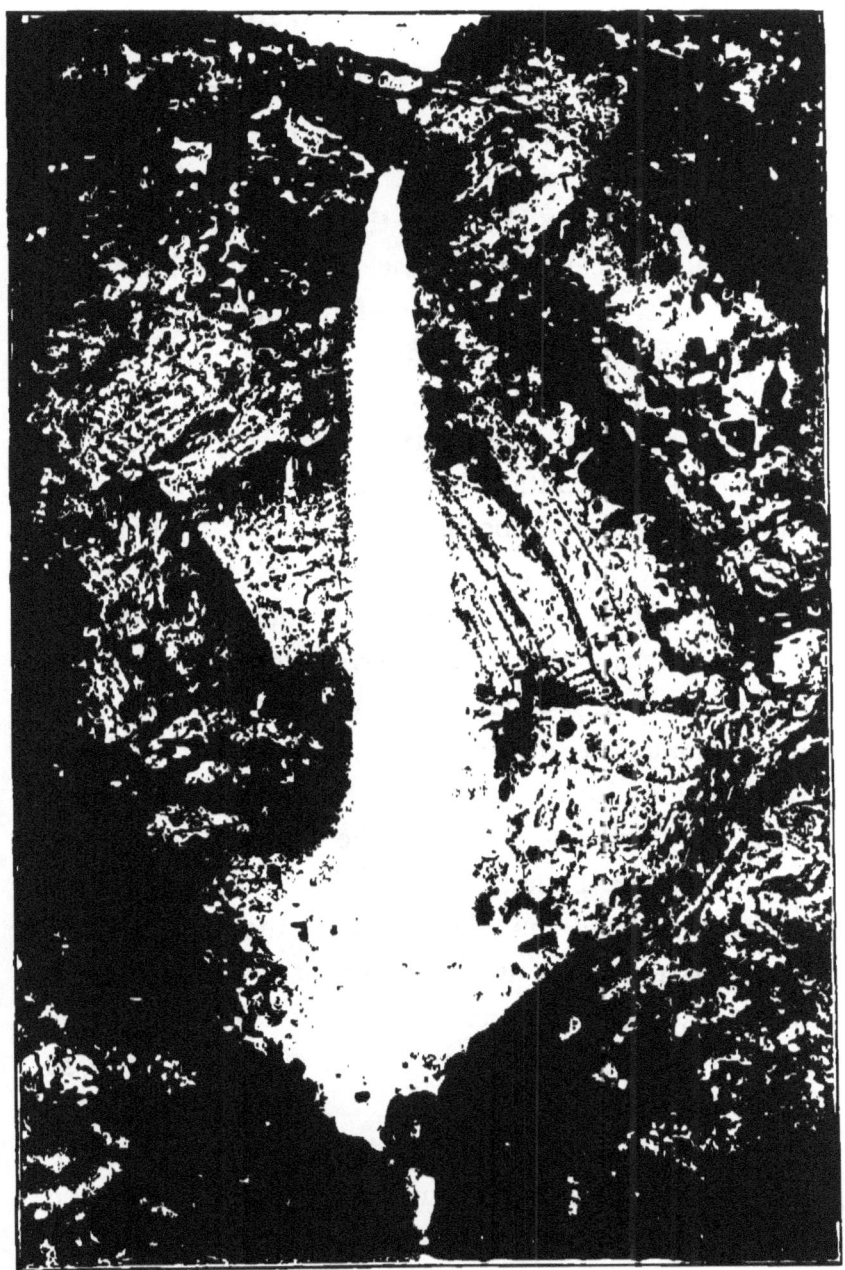

FIG. 1.—Cascade of Arpenaz.

fore the Jurassic period, for rolled pebbles of it occur in that rock. On the other hand, the fold on the summit of the Windgälle contains Eocene strata. The origin of the Porphyry then is earlier than the Jurassic; the elevation of the mountain is later than the Eocene. It is clear, therefore, that the Porphyry had nothing whatever to do with the origin of the Windgälle mountain.

The igneous rocks have moreover produced no effect on the strata which now rest on them. If, however, they had been intruded in a molten condition, they must have modified the rocks for some distance around. It is evident therefore that the igneous rocks had cooled down before the overlying strata were deposited. The elevation of the Alps only commenced in the Tertiary period, but we know that the Granite of the southern Alps is, for the most part, pre-Carboniferous, that the Porphyry of Botzen belongs to the Permian period, the younger Porphyry to the Trias, and that the Gneiss of the central range of the eastern Alps is still older; it is evident then, that these plutonic rocks can have taken no active part in the upheaval of the Alps, which occurred so much later.

We may, indeed, lay it down as a general proposition that folded mountains are not due to volcanic action. When the two are associated, as in the Andes, the volcanoes are due to the folding and crushing, not the folding to the volcanoes.

The Alps then have not been forced up from below, but thrown into folds by lateral pressure. This view was first suggested by De Saussure, worked out in fuller detail by Sir Henry De La Bêche in 1846, and recently developed by Ball, Suess, and especially by Heim.*

Moreover, as the following sections show, (Figs. 1, 5, 23-26) we have every gradation from the simple undulations of the Jura (Fig. 5) to the complicated folds of the Alps (Figs. 1, 23-26).

But why are the surface strata thus thrown into folds? When an apple dries and shrivels in winter, the surface becomes covered with ridges. Or again, if we place some sheets of paper between two weights on a table, and then bring the weights nearer together, the paper will be crumpled up.

In the same way let us take a section of the Earth's surface AB (Fig. 2) and suppose that, by the gradual cooling and consequent contraction of the mass, AB sinks to $A^1 B^1$, then to $A^2 B^2$, and

---

* See especially Heim's great work, *Untersuchungen ü. d. Mechanismus d. Gebirgsbildung.* I ought perhaps, however, to add that this view is not universally accepted.

THE ORIGIN OF MOUNTAINS. 59

finally to $A^3 B^3$. Of course if the cooling of the surface and of the deeper portion were the same, then the strata between A and B would themselves contract, and might consequently still form a regular curve between A^3 and B^3. As a matter of fact, however, the strata at the surface of our globe have long since approached a constant temperature. Under these circumstances there would be no contraction of the strata between A and B corresponding to that

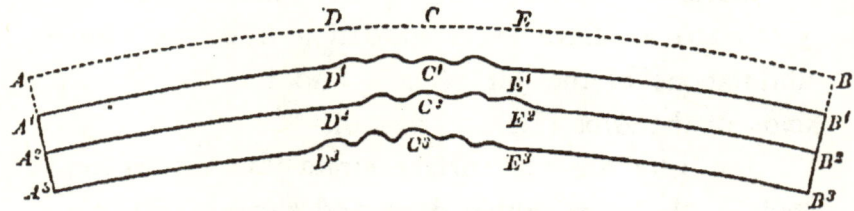

FIG. 2.—Diagram in explanation of folded mountains.

in the interior, and consequently they could not lie flat between A^3 and B^3, but must be thrown into folds, commencing along any line of least resistance. Sometimes, indeed, the strata are completely inverted, and in other cases they have been squeezed for miles out of their original position. "The great mountain ranges," says Geikie, "may be looked upon as the crests of the great waves into which the crust of the Earth has been thrown." Sir James Hall illustrated the origin of folds very simply (Fig. 3) by

Fig. 4.—Showing the artificial folds produced in a series of layers of clay on indiarubber, according to an experiment by Prof. A. Favre.

placing layers of cloth under a weight, and then compressing the two sides, and more complete experiments have since been made by Favre, Ruskin and Cadell.

Fig. 4 shows the result of one of Favre's experiments, in which he used the contraction

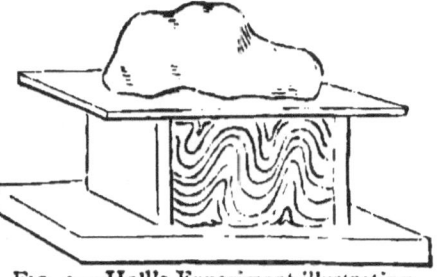

Fig. 3.—Hall's Experiment illustrating Contortion.

of an indiarubber band to produce the folds.

The shortening of the Jura amounts to about one-fifteenth. The strata between Basle and the St. Gotthard, a distance of 130 miles, would, if horizontal, occupy 200 miles. Heim estimates the total compression of the

Alps at a minimum of 120,000 metres.* The original breadth of the strata forming the Aarmassif was at least double the present, and the same may be said of the central range. The Appalachians are calculated to be compressed from 150 miles to 65.

It very seldom happens that such a range of mountains consists of a single fold. There are generally several, one being as a rule formed first, and others outwards in succession. In both the Alps and the Jura, the southern folds are the oldest. In Central America, again, there are several longitudinal ranges, and the volcanoes are generally situated on cross lines of fracture, so that they are in rows, at right angles to the general direction of the mountains, and in almost every case the outer crater, or that towards the Pacific, is the only one now active.

A glance at any good map of the Jura will show a succession of ridges running parallel to one another in a slightly curved line from south-west to northeast. That these ridges are due to folds of the Earth's surface is clear from the following figure (Fig. 5) in Jaccard's work on the Geology of the Jura, showing a section from Brenets due South to Neuchâtel by Le Locle. These folds are com-

* *Mechanismus d. Gebirgsbildung*, v. 2.

paratively slight and the hills of no great height. In the Alps the strata are much more violently dislocated and folded.

The mountains seem so high that we are apt to exaggerate the relative elevation. The following figure (Fig. 6) by Rütimeyer gives the outline of the Alps from Basle to near Milan. This section is only intended to indicate the relative height, and is supposed to follow the line of one of the great valleys. Even so, however, it ought to have shown the sudden dip to the south of the main ridge.

The folded structure throws light on the curious fact that there are much fewer faults in Switzerland than in such a region as, for instance, that of our coal fields.

In folded districts the contortions are often so great that if we could not follow every step they would certainly be regarded as incredible. Previous folds are themselves in some cases refolded, and in others the lateral pressure has not only raised the strata into a vertical position, as for instance the Chalk and Tertiary sands of Alum Bay in the Isle of Wight, but has in some cases pushed the folds for miles, and has even thrown them over, so that the sequence is inverted, and the more ancient lie over the more recent strata in reverse order. As

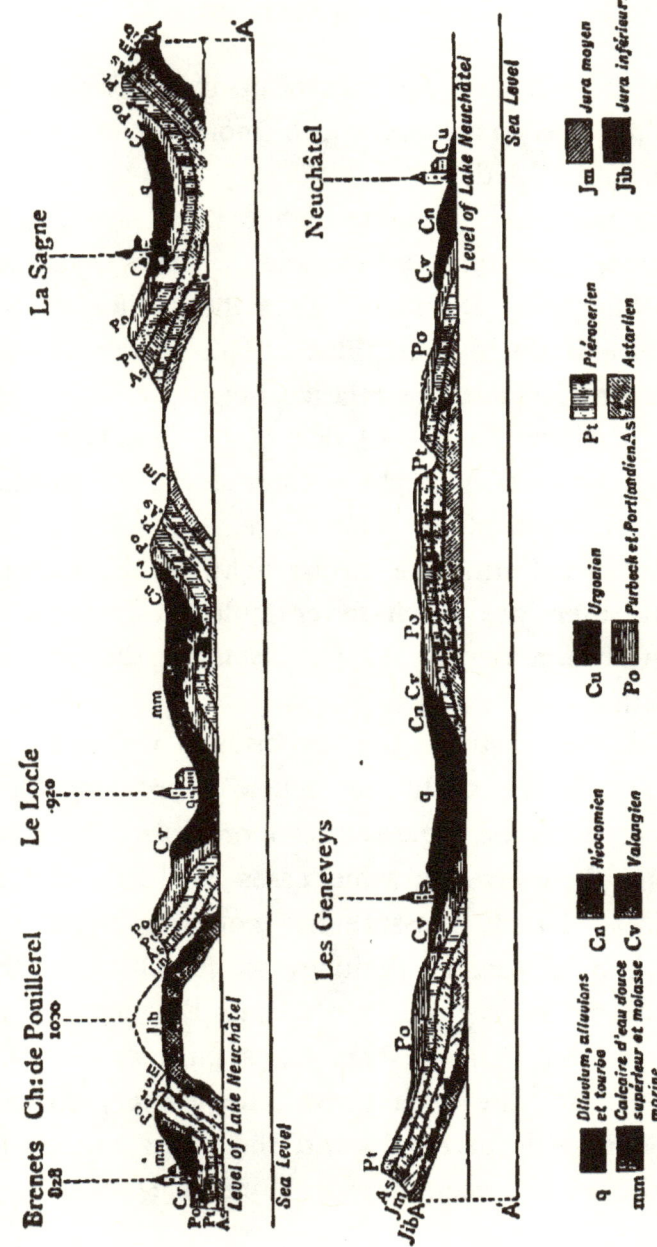

Fig. 5.—Section across the Jura from Brenets to Neuchâtel.

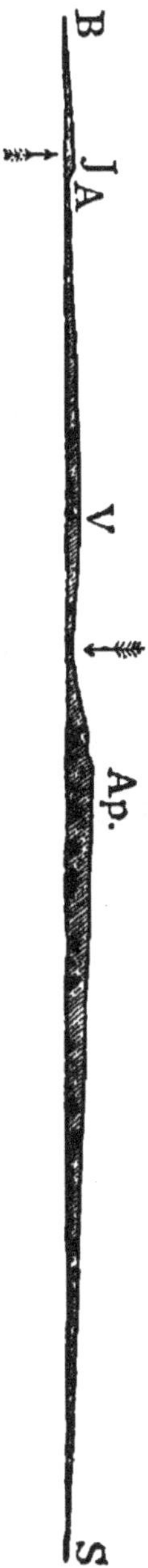

the cooling, and consequent contraction of the Earth, is a continuous process, it follows that mountain ranges are of very different ages; and, as the summits are continually crumbling down, and rain and rivers carry away the debris, the mountain ranges are continually losing height. Our Welsh hills, though comparatively so small, are venerable from their immense antiquity, being far older, for instance, than the Vosges, which themselves, however, were in existence while the strata now forming the Alps were still being deposited at the bottom of the Ocean. But though the Alps are from this point of view so recent, it is probable that the amount which has been removed is almost as great as that which still remains. They will, however, if no fresh elevation takes place, be still further reduced, until nothing but the mere stumps remain. What an enormous amount of denudation has already taken place is shown for instance in Fig. 7, representing the mountain of Tremettaz near the valley of the Rhone, between

FIG. 6.—Section from Basle across the Alps to Senago, north-west of Milan. *B*, Basle. *J*, Jura. *A*, Aar, near Olten. *Ap*, Alps. *S*, Senago. Inclination from Basle to summit of Jura, 1° 37´; from Basle to the summit of the Alps, 1° 43´.

THE ORIGIN OF MOUNTAINS.

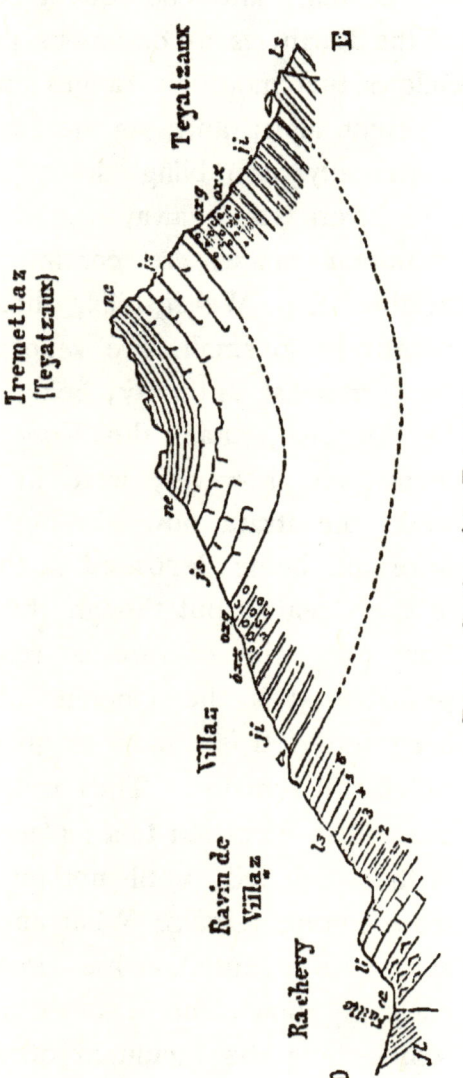

Fig. 7.—Section of the Tremettaz.

the Niremont and the valley of the Sarine, where it is evident, not only that the strata have been cut off, but that what is now the top of the mountain was once the bottom of a valley.

The edges of strata which appear at the surface of the ground are termed their "Outcrop." Sometimes they are horizontal, but if not, the inclination is termed their "Dip" (Fig. 8, B). A horizontal line drawn at a right angle to the Dip is called the "Strike" (Fig. 8, A) of the rocks. If the surface of the ground is level this will coincide with the outcrop. In a mountainous district such as Switzerland this is however rarely the case.

Where strata have been bent, as in Fig. 9, it is called a monoclinal fold. Where the subterranean forces have ruptured the strata and pushed the one side of the crack more or

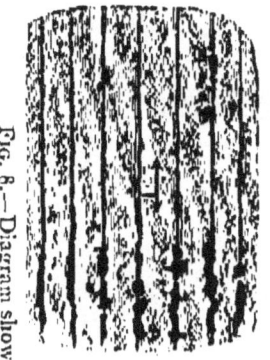
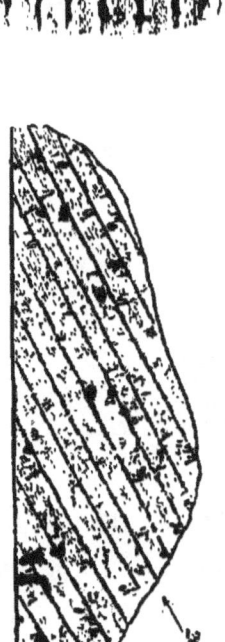

Fig. 8.—Diagram showing (A) the "Strike" and (B) the "Dip" of strata.

less upwards or downwards (Fig. 10), it is termed a fault.

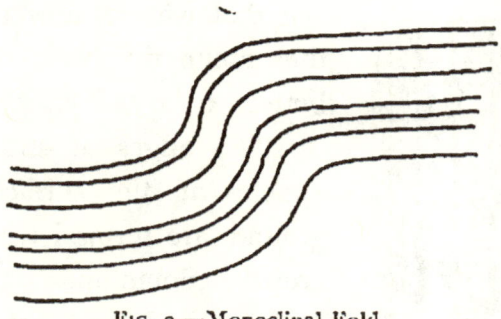

Fig. 9.—Monoclinal Fold.

Faults may be small, and the difference of height between the two sides only a few inches. On the other hand, some are immense. In the case of one

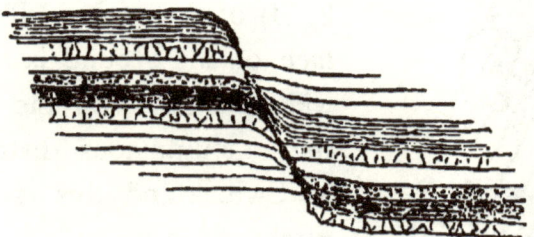

Fig. 10.—A Fault.

great fault described by Ramsay, the difference is no less than 29,000 feet, and yet so complete has been the denudation that the surface shows no evidence of it, and one may stand with a foot on each side, unconscious of the fact that the stratum under the one represents a geological horizon so much above that under the other.

When the strata bent somewhat before the fracture we have a fold-fault (Fig. 11).

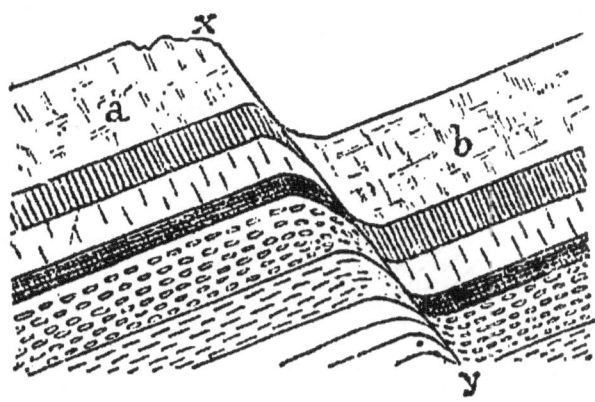

FIG. 11. Line of Fault at the upper displaced bed. The beds are bent near the fault by the strain in slipping.

Where a fold is much compressed the limbs would become thinner and thinner (Fig. 12), while the strata in the arch and the trough would be compressed and consequently widened.

FIG. 12.—An Inclined Fold.

When the arch A, instead of being upright is

thrust to one side, it is said to be inclined or recumbent (Fig. 12).

Where strata are thrown into folds the convex portion is termed an anticlinal (Fig. 14, *A*) and the concave a synclinal (Fig. 14, *B*). The same terms are applicable when the surface has been planed down so that the strata would dip as in Fig. 13.

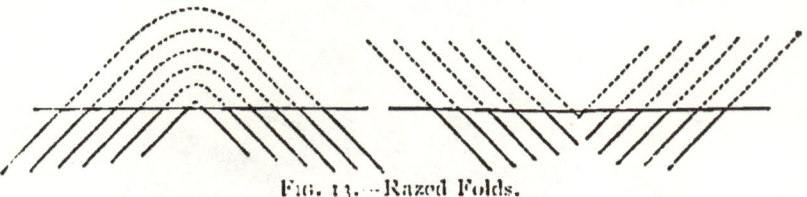

Fig. 13.—Razed Folds.

The inner strata of any fold are called the core, those of an anticlinal (Fig. 14, *A*) being called the arch core, those of a synclinal (Fig. 14, *B*) the trough core.

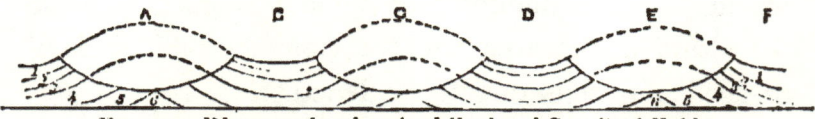

Fig. 14.—Diagram showing Anticlinal and Synclinal Folds.

It is obvious of course that when strata are thrown into such folds, they will, if strained too much, give way at the summit. Before doing so, however, they are stretched and consequently loosened, while on the other hand the strata at the bottom of the fold are compressed; the former, therefore, are rendered more susceptible of disintegration, the latter on the contrary acquire greater powers of resistance.

The above diagram, Fig. 14, represents six strata (1-6) supposed to be originally of approximately equal hardness, but which, after being thrown into undulations, are rendered more compact in the hollows and less so in the ridges. Denudation will then act more effectively at *A, C, E,* than at *B, D, F,* and when it has acted long enough the surface will be shown by the stronger line. This will be still more rapidly the case if some of the strata are softer than others. Where they are brought up to the surface erosion will of course act with special effect. Hence it often happens that hills have become valleys, and what were at first the valleys have become mountain tops. As an illustration of the former I may mention the valley of the Tinière (Fig. 98, vol. II. p. 87); of the latter, the Tremettaz (Fig. 7) or the Glärnisch.

In other cases where the summit is not at the very base of the trough, the edges of some stratum rather harder than the rest, project as two more or less pointed peaks, leaving a saddle-shaped depression in the centre.

Highly inclined strata are often worn away so as

to form a kind of wall, sometimes so thin that it is actually pierced by a natural hole, as for instance the Martinsloch above Elm, in Glarus. There is another of these orifices near the summit of the Pilatus, one in the Marchzahn, a mountain of the Gastlose chain, and another in the Piz Aela, also known for that reason as Piz Forate, between the Albula and the Oberhalbstein Rhine.*

When we look at these abrupt folds and complicated contortions, the first impression is that they must have been produced before the rocks had solidified. This, however, is not so. They could not indeed have been formed except under pressure. We must remember that these rocks, though they are now at or near the surface, must have been formerly at a great depth, and where the pressure would be tremendous. Even in tunnels, which of course are comparatively near the surface, it is sometimes found necessary to strengthen and support the walls which would otherwise be crushed in. The roadways in coal-mines are often forced up, especially where two passages meet. This indeed is so common that it is known as the "creeps." In deep tunnels it has not unfrequently happened that when strata have been uncovered they have suddenly bent

* Theobald's Graubünden, *Beitr. z. Geol. Karte d. Schw.*, II.

and cracked, which shows that they were under great lateral pressure. Yet the deepest mine only reaches 800 metres.

Treska* has shown by direct experiment that the most solid bodies, lead, tin, silver, copper, and even steel, will give way and "flow" under a pressure of 50,000 kilograms per square centimetre. Moreover, there is direct and conclusive evidence that the Swiss rocks were folded after solidification. In many cases contorted rocks contain veins (Fig. 16) which are in fact cracks filled up with calcite, etc. Such fine fissures, however, can only occur in hard rock. Again the Eocene contains rolled pebbles of Gneiss, Lias, Jurassic, etc., which must therefore have become hard and firm before the Eocene period,** while the folding did not occur till afterwards. It is clear therefore that when the folding took place the rocks were already solidified. No doubt, however, the folding was a very slow process. It took place, and could only take place, deep down, far below the surface, under enormous pressure, and where the material was perhaps rendered somewhat more plastic by heat. In the later and higher rocks we find compression with fracture, in the earlier and lower rocks

* *Comptes Rendus*, 1874.
** Heim, *Mech. d. Gebirgsb.*, vol. II.

compression with folding. Whenever we find a fold we may be sure that, when formed, it was deep down, far below the surface.

In fact folds and fractures are the two means by which the interior strains adjust themselves. They replace one another, and in the marvellously folded districts of the Alps faults are comparatively few, though it must not be supposed that they do not occur. The nature of the rock has little influence on the great primary folds, but the character of the minor secondary folds depends much upon it.

Fig. 15 represents a piece of contorted mica schist, and it will be seen that the folds are a miniature of those to which on a great scale our mountains are due.

Many of the following figures give an idea of the remarkable folds and crumpling which the strata have undergone, so much so that they have been compared to a handful of ribbons thrown on to the ground.

It is obvious that before strata could be thrown into contortions such as these, they must have been subjected to tremendous pressure. They have consequently been much altered, and the fossils have been compressed, contorted, crushed, ground, and partly, or in many cases entirely, obliterated.

In parts of the great Glarus fold (see vol. II. p. 64) the Hochgebirgskalk is reduced from a thickness of 450

Fig. 15.—Hand Specimen of Contorted Mica Schist.

to a few metres.* In other cases certain formations have been completely squeezed out. We must not

* Heim, *Mech. d. Gebirgsb.*, vol. II.

therefore infer, from the absence of a given stratum in such cases, that it never existed.

In many cases the rock is broken up into flat or more or less lenticular pieces, which have been squeezed over one another so that their surfaces have been rendered smooth and glistening. Such surfaces are known as slickensides. This process has sometimes been so intense and so general that hardly a piece can be found which does not present such a polished surface. The particles of stone which now touch were once far apart, others which are now at a distance once lay close together. The cracks, movements, and friction which result in such a structure must from time to time produce sounds, and the mysterious subterranean noises sometimes heard are perhaps thus produced.

Fig. 16 represents a section of Röthidolomite, and it will be observed that, as we should expect theoretically (see also Fig. 12, p. 68), the strata are thinnest in the limbs, where they are squeezed out, and broader in the arches. This is visible in great mountain folds, as well as in hand specimens.

In the part of the curve where the effect of the force is to draw out the strata, they will as shown above, if capable of giving way, become thinner. If however they are not plastic they must crack, the combined width of the cracks affording the additional space. Fig. 17 represents a fragment of Verrucano thus drawn out.

Fig. 16.—Section of Röthidolomite.

In many cases fossils are compressed or torn, but still distinguishable. Fig. 18 represents Belem-

Fig. 17.—Piece of Stretched Verrucano.

nites thus compressed and torn; but in all these

cases the extension or tearing is due, not to a general extension of the rock, but to lateral thrust.

Fig. 19 represents a piece of nummulitic limestone in which the rock has not only been fractured along the lines *a b,* but two sides of the vein *a* have been evidently displaced. At a later date another fracture has taken place along the line *c d.*

Some rocks have been so kneaded and ground together that in many places it is rare to find a cubic millimetre next its original neighbours.* In many places fragments and wedges of one formation have been forced into another.

In the Tertiary slates of the Sernfthal at Platten-

* Heim, *Mech. d. Gebirgsb.*, vol. I.

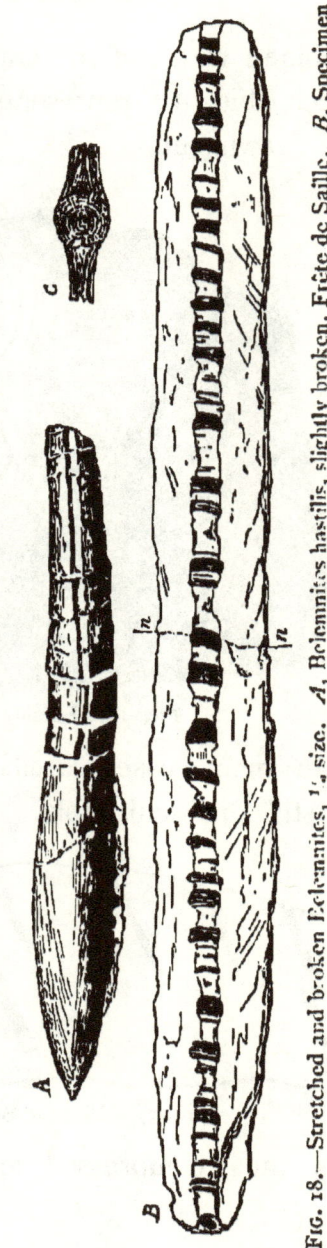

FIG. 18.—Stretched and broken Belemnites, ½ size. *A*, Belemnites hastilis, slightly broken, Frête de Saille. *B*, Specimen much drawn out. *C*, Section at *n*.

berg near Matt are well-preserved remains of fish belonging to the genus Lepidotus. Agassiz thought he could distinguish, and described, six species, but Wettstein has shown that they all belong to one and the same, and that the differences of form are merely due to the position in which the specimens

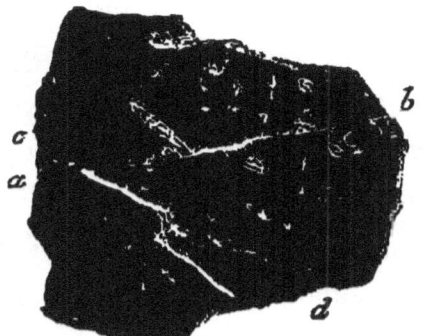

Fig. 19.—Fragment of Nummulitic Limestone.

happened to lie with reference to the direction of pressure.

In many cases the pressure has produced "cleavage," and turned the rocks into slate,* so that they split into more or less perfect plates or films. The direction of cleavage is quite independent of

---

* English geologists apply the term "shale" to rocks which split along the laminæ of original deposition, and which are comparatively soft and destructable, and "slate" to those where the lamination is due to cleavage. Continental geologists generally include shales and slates under the same name.

the stratification, which it may cross at any angle. Heim distinguishes three forms of cleavage. Firstly, that due to the formation of slickensides as just described (*ante*, p. 75). The second kind of cleavage is due to the minute particles in the rock being flattened by, and arranged at right angles to, the pressure, as shown in Figs. 20 and 21.*

The third is produced by all the laminæ or elongated particles being arranged by the pressure in lines of least resistance, so that they are forced to lie parallel to one another.

It is, however, by no means always easy, especially in the crystalline rocks, to distinguish cleavage from stratification. The structure of the rock, which forms the base of the Windgälle, and which Heim regards as partly stratification, is considered by some geologists to be all cleavage.

The fact that cleavage has been produced by pressure was first demonstrated by Sharpe, and afterwards with additional evidence by Sorby and Tyndall. In fact, under great pressure solid rock behaves very much like ice in a glacier.

Cleavage and folding are both due to the same

* Geikie, *Text-book of Geology*.

cause. They have arisen simultaneously, and are

Fig. 20. — Section of a fragment of argillaceous rock.

different manifestations of the same mechanical action.

Fig. 21. Section of a similar rock which has been compressed, and in which cleavage structure has been developed.

CHAPTER III.

THE MOUNTAINS OF SWITZERLAND.

Erst dann haben wir ein Gebirge erkannt, wenn sein Inneres durchsichtig wie Glas vor unserem geistigen Auge erscheint.
THEOBALD.

We do not really know a mountain until its interior is to our mental eye as clear as crystal.

THE Swiss mountains, as indicated in the preceding chapter, are now considered to be due, not to upheaval from below, but to lateral pressure.

This acted from the south-east to north-west, and took place at a comparatively recent period, mainly however after the end of the Eocene period. There are good grounds for supposing that a former range occupied the site of the present Alps at an early period, and the Carboniferous strata show considerable folds (Fig. 22), over which the Permian and more recent strata were deposited.

The Carboniferous Puddingstone of Valorsine, which contains well-rounded pebbles and boulders, shows that there must have been mountains and rapid rivers at this period. These ancient mountains, however, were removed by denudation, and the whole country sunk below the Sea. Between the Eocene and the Miocene was a second period of disturbance, and all the strata, including the Eocene, were folded conformably together.* The main elevation of the Alps was, however, between the Miocene

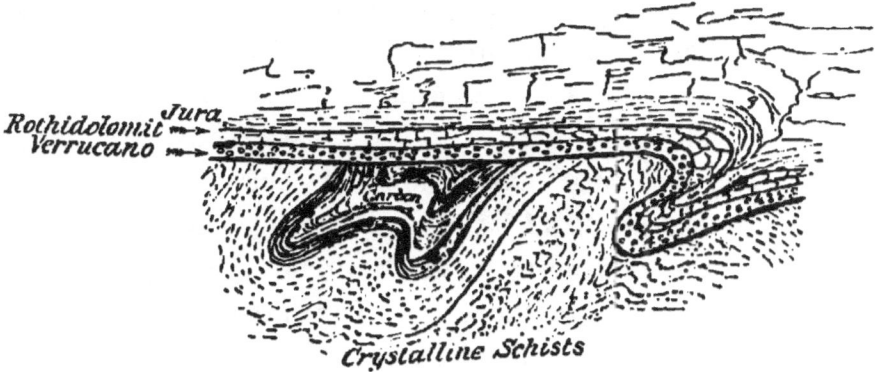

FIG. 22.—Carboniferous Folds on the Biferten Grat.

and the Glacial periods. Miocene strata attain in the Rigi a height of 6000 feet. By this much at least then the Alps must have been raised since the close of this comparatively recent period.

"It is strange to reflect," says Geikie, "that the enduring materials out of which so many of the mountains, cliffs, and pinnacles of the Alps have been formed are of no higher geological antiquity than the

* Heim, *Mech. d. Gebirgsb.*, vol. I.

London Clay and other soft Eocene deposits of the South of England."*

Unfortunately we seldom see a map, except on quite a small scale, of the whole Alps. We have separate maps of France, of Switzerland, of Italy, and of the Austrian dominions. But to get a good general idea of the whole Alps, we require not only Switzerland, but parts of France, Italy, and Austria. If we have such a map before us we see that, with many minor irregularities, the Alps are formed on a definite plan.

The principal axis follows a curved line encircling the North of Italy; commencing with a direction almost due north in the Maritime Alps, sweeping round gradually to the east. The direction appears to have been determined by the pre-existing Central Plateau of France and the Black Forest, which probably formed a continuous barrier before the subsidence of the Rhine valley. They are in fact ancient pillars, far older than the Alps, and Switzerland has been thrown into waves or folds by compression against these great buttresses.

"'To account for the conformation of the Alps," says Tyndall, "and of mountain regions generally, constitutes one of the most interesting problems of

---

* Geikie's *Text-book of Geology*.

the present day. Two hypotheses have been advanced, which may be respectively named the hypothesis of *fracture* and the hypothesis of *erosion*. Those who adopt the former maintain that the forces by which the Alps were elevated produced fissures in the earth's crust, and that the valleys of the Alps are the tracks of these fissures. Those who hold the latter hypothesis maintain that the valleys have been cut out by the action of ice and water, the mountains themselves being the residual forms of this grand sculpture. To the erosive action here indicated must be added that due to the atmosphere (the severance and detachment of rocks by rain and force), as affecting the forms of the more exposed and elevated peaks." *

This was written thirty years ago and has been confirmed by the subsequent researches of geologists. While the folding referred to in the last chapter has elevated the ranges and determined the position of many of the Swiss valleys, "fracture" has played but a subordinate part, and to denudation and erosion, as Tyndall himself always maintained, the present conformation of the country is mainly due.

* Tyndall, "Conformation of the Alps," *Philosophical Mag.*, Oct. 1864. See also Scrope, "On the Origin of Valleys," *Geol. Mag.* 1866.

Switzerland is divided roughly into equal parts by four great rivers,—the Rhine, the Rhone, the Reuss, and the Ticino. These four rivers rise on the same great "central massif." The valleys are not, however, of the same character. The Rhine-Rhone valley from Martigny to Chur is a "geotectonic" valley; its direction coincides with the direction or "strike" of the strata, and it was originally determined by a great fold in the strata.

The Reuss and Ticino valleys (except the upper part of the Reuss in the Urserenthal, which is in fact a part of the Rhone-Rhine valley and the upper part of the Ticino in the Val Bedretto, which is also a longitudinal valley) are transverse; they cross the strata approximately at right angles, and consequently the rocks on the two sides are the same. They are entirely or almost entirely due to erosion.

In the Jura, where the foldings are comparatively gentle and the denudation has been much less, the present configuration of the surface follows more closely the elevations and depressions due to geological changes (see Fig. 5).

In the Alps the case is different, and the denudation has so far advanced that we can at first sight trace but little relation between the valleys, as indicated by the river courses and the mountain chains, and the geological structure of the country. There are many cases of anticlinal valleys; that is to say, of valleys (see *ante,* p. 70) which run along what was at one time the summit of an arch, as, for instance, the Maderanerthal (Fig. 24) and the Val de la Tinière (Fig. 98).

In other cases a piece is cut off from the rest of the massif to which it belongs, as, for instance, the Frusthorn from the Albula massif by the Valserthal.

There are others where a mountain, or range of mountains, occupies the line of a former valley. This is the case for instance with the mountain ridge which runs between the Rhine and the upper Linth from the Kistenpass at the head of the Limmerbach to the south of the Limmern Glacier, by the Bifertenstock to Piz Urlaun and Stock Pintga or the Stockgron.* This range of mountains occupies the site of an original valley, but no doubt from the greater hardness of the rock and its position it has offered a more successful resistance to attack; while the original mountains have been washed away.

In this way some at anyrate of the transverse ranges have, as it were, been carved out. Thus the Safienthal—the valley of the Glenner which falls into

* Heim, *Beitr. z. Geol. K. d. Schw.*, L. xxv.

the Rhine at Ilanz—is bounded by ranges approximately at right angles to the main direction of the mountains. That on the left of the valley culminates in the Piz Ricin, Crap Grisch, Weissensteinhorn, and Bärenhorn. In favourable light it can easily be seen from the opposite side of the valley, that the streams have cut out the valleys and are thus the cause of the mountains. This is a particularly clear illustration, because the strata are uniform along the whole line, so that the structure is not complicated by the presence of rocks of different character and hardness.

Indeed if we compare together two maps, in one of which the principal chains of mountains, and in the other the main river valleys, are brought out most prominently, they look at first sight so different that we should hardly suppose them to represent the same district.* It is evident therefore that the main agent which has determined the longitudinal valleys is not that which has given rise to the mountain summits. The courses of the rivers, though there have, as we shall see, been many minor changes, and exceptions due to other causes, still were determined by the folds into which the surface was thrown; while

* Heim, *Mech. d. Gebirgsb.*, vol. I.

the present mountain summits are mainly the result of erosion and denudation.

We will now consider the evidence which leads to the conclusion that the fossiliferous strata formerly extended over the Central chain of the Alps. It is a common error to suppose that the limits of geological strata are those which are now shown on the map. It requires little reflection however to show that this was not so. In the abyssal depths of the ocean deposit is portentously slow, and a long period would be represented by only a few inches of rock. Moreover, though a marine formation proves the existence of sea, the absence of a marine formation does not prove the existence of land. Strata may and often have been entirely removed. Our Cretaceous deposits, for instance, once extended far beyond their present limits. The same was the case with the Secondary deposits of Switzerland from the Trias to the Eocene. They extended completely over the Central mountains. If these mountains had been then in existence and the Secondary strata had been deposited round them, we should find evidence of shore deposits, with remains of animals and seaweeds such as live in shallow waters and near land. This is however not the case; we find no pebble beds such as would be the case near a shore, no gravels with pebbles of granite, gneiss, or crystalline schists, but deep-sea deposits of fine sediment evidently formed at some distance from land. In the Triassic period there seems to have been a barrier between the Eastern and Western Alps, but subsequently the conditions must have been very similar, and the southern shores of the Jurassic Sea were perhaps far away in Africa.*

Even the Eocene deposits show no evidence of a shore where the Alps now rise above them.

We have other proofs that the central chains were formerly covered by other strata. For instance, the Puddingstone of Valorsine at the head of the Chamouni valley, which belongs to the Carboniferous period, contains no granite or porphyry pebbles. The granite and porphyry strata of the district must therefore at that period have been protected by a covering of other rocks which have been since stripped off.

It is also significant that the pebbles of the Miocene Nagelflue which come from the neighbourhood are mainly of Eocene age. Neither the Crystalline rocks nor the older Secondary strata seem to have been then as yet uncovered.** There are indeed

* Heim, *Mech. d. Gebirgsb.*, vol. II.; Baltzer, *Beitr. z Geol. K. d. Schw.*, L. XXIV.
** Heim, *Mech. d. Gebirgsb.*, vol. II.

crystalline and Triassic pebbles in the Nagelflue, for instance, of the Rigi, but they do not belong to rocks found in the valley of the Reuss or on the St. Gotthard. They resemble those of Lugano, Bormio, the Julier, and other districts far away to the southeast.

We are not however dependent on these arguments alone, conclusive as they are. Remains of Secondary strata occur here and there in the Central district, and these are not fragments torn away from one another, but parts of a formerly continuous sheet, which have been preserved in consequence of being protected in the hollows of deep folds. That the Secondary strata were once continuous over the Central chain is well shown in the following section (Fig. 23) drawn from the Rhone to the Averserthal and cutting the Binnenthal, Val Antigorio, Val Bavena, Val Maggia, Val Ticino, Val Blegno, Val Misocco and Val St. Giacome. It will be seen that all these valleys are primarily due to great folds, and that in each case we find at the bottom of the valleys remains of the Secondary strata nipped in between the Crystalline rocks.

Fig. 24 shows a section after Heim, from the Weisstock across the Windgälle to the Maderanerthal. It is obvious that the valleys are due mainly to erosion, that the Maderaner valley has been cut out of the Crystalline rocks, *s,* and was once covered by the Jurassic strata *j*, which must have formerly passed in a great arch over what is now the valley.

Again it is clear (Fig. 25) that a great thickness of Crystalline rock has been removed from the summit of Mont Blanc. No doubt (see *ante,* p. 28) many thousand feet had been removed before the deposition of the Secondary strata. But even since its elevation the amount of erosion of the Granite itself has been considerable. How much we do not know, but 500 metres would probably be a moderate estimate. To this must be added the Crystalline Schists, say 1000 metres, and the Sedimentary rocks, which from what we know of their thickness elsewhere cannot be taken at less than 3000 metres. This therefore gives 4500 metres, or say 14,000 feet, which

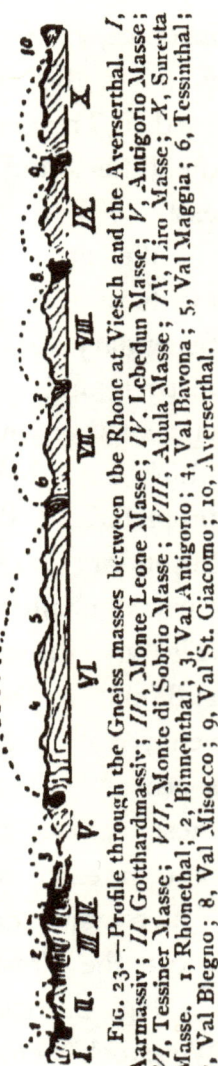

FIG. 23.—Profile through the Gneiss masses between the Rhone at Viesch and the Averserthal. *I,* Aarmassiv; *II,* Gotthardmassiv; *III,* Monte Leone Masse; *IV,* Lebedun Masse; *V,* Antigorio Masse; *VI,* Tessiner Masse; *VII,* Monte di Sobrio Masse; *VIII,* Adula Masse; *IX,* Liro Masse; *X,* Suretta Masse. 1, Rhonethal; 2, Binnenthal; 3, Val Antigorio; 4, Val Bavona; 5, Val Maggia; 6, Tessinthal; 7, Val Blegno; 8, Val Misocco; 9, Val St. Giacomo; 10, Averserthal.

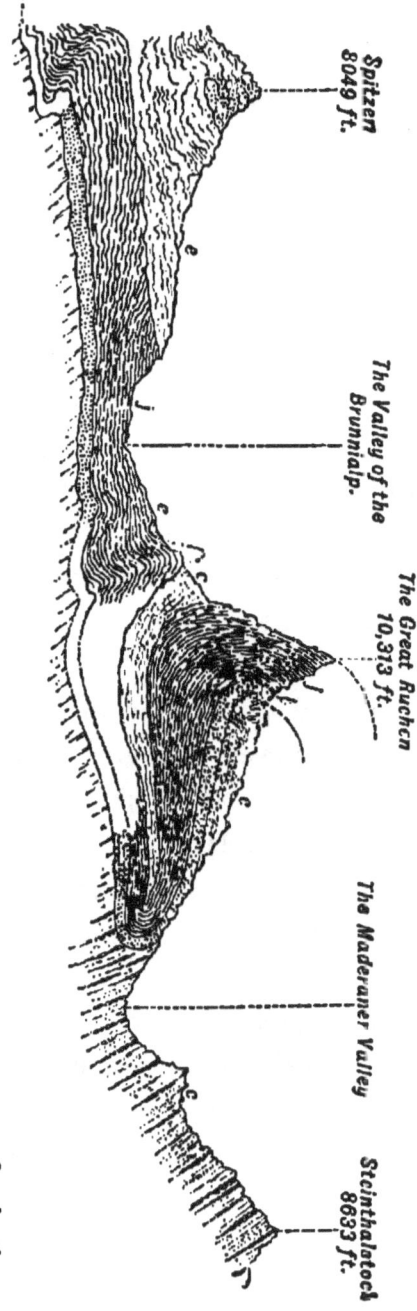

Fig. 24.—Section from the Weisstock across the Windgälle to the Maderanerthal.

erosion and denudation have stripped from the summits of the mountains! Fig. 26 gives a section across the Alps, and it will be seen that the section across the St. Gotthard substantially resembles that of Mont Blanc.

Surprising, and even almost incredible, as this may at first sight appear, it becomes less difficult to believe when we remember that not only the great Miocene gravel beds which form the Central plain of Switzerland, but much of the deposits which occupy the valleys of the Rhine, Po, Rhone, Reuss, Inn, and Danube — the alluvium which forms the plains of Lombardy, of Germany, of Belgium, Holland, and of South-east France consists of materials washed down from the Swiss mountains.

It is calculated that at the present rate of erosion the Mississippi removes one foot of material from its drainage area in 6000 years, the Ganges above Ghazipur in 800, the Hoangho in 1460, the Rhone in 1500, the Danube in 6800, the Po in 750. Probably therefore we may take the case of the Rhone as approximately an average, and this gives us, if not a measure, at anyrate a vivid idea of the immense length of time which must have elapsed.

The great plain shows comparatively gentle elevations, which become more marked in the "Prealps,"

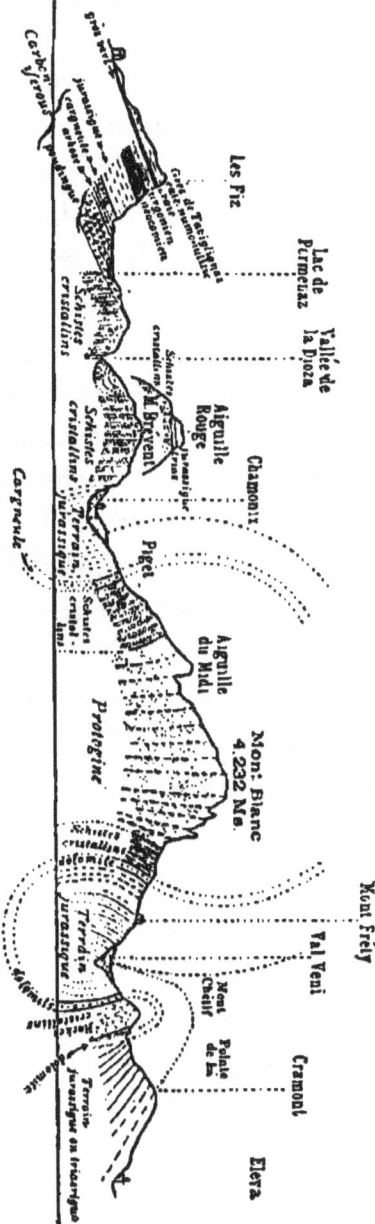

FIG. 25.—Section across the Mont Blanc range.

while the inner chains are thrown into the most extreme contortions. In some cases the result of compression has been to push certain strata bodily over others. Such overthrusts also greatly tend to render the relief of the surface independent of the tectonic structure. If there were no overthrusts, if the arches had been flatter and the troughs broader, the causes which have led to the present configuration of the surface would have been much clearer.

The main ranges then are due to compression and folding, the peaks to erosion, and the three main factors in determining the physical geography of Switzerland, have been compression, folding, and denudation.

The present configuration of the surface is indeed mainly the result of denudation, which has produced the greatest effect in the Central portions of the chain. It is probable that the amount which has been

FIG. 26.—Section across the Alps.

removed is nearly equal to that which still remains,* and it is certain that not a fragment of the original surface is still in existence, though it must not be inferred that the mountains were at any time so much higher, as elevation and denudation went on together.

The country is now, however, so well mapped that if changes are still going on they must ere long show themselves. It is probable on mechanical and geological grounds that the southern chains were formed first, and the northern ones afterwards in order of succession. It has been shown that the Secondary strata originally covered the whole area, and their removal from the Central massives, except in the deeper folds, is strong evidence of their great age. This leads us to the consideration whether changes of level are still taking place. There are some reasons for doubting whether they have altogether ceased, but as yet we have no absolute proof.

Slight earthquakes are common in Switzerland; more than 1000 have been recorded during the last 150 years, and no doubt many more have passed unnoticed. This appears to indicate that the forces which have raised the Alps are perhaps not entirely

---

* Heim, *Beitr. z. Geol. K. d. Schw.*, L. xxv.

spent, and that slow movements may be still in progress along the flanks of the mountains.*

Many of these earthquakes are very local and as a rule not deep seated, at a depth of not more than from 15,000 to 20,000 metres.

Even however in the Central Alps there is some evidence of present strain. When the tunnels were being pierced for the St. Gotthard line, and especially the Wattinger tunnel near Wasen, slight explosions were often heard, and blocks of rock were thrown down on the workmen. These generally came from the roof, but sometimes from the sides, and eventually it was found necessary to case the interior of the tunnel.** These phenomena, however, may have been only due to the great pressure.

The American geologists, and especially Dana, have pointed out that folded mountains are not as a rule symmetrical but one-sided. Suess *** has extended this to Switzerland, and indeed to folded mountains generally. It is remarkable that in all the European mountain systems—the Alps, Appennines, Jura, Carpathians, Hungarian Mountains, etc., the outer side of the curve presents a succession of folds

---

* Heim, *Mech. d. Gebirgsb.*, vol. II.
** Baltzer, *Beitr. z. Geol. K. d. Schw.*, L. XXIV.
*** *Das Antlitz der Erde.*

which gradually diminish in intensity, while the inner side terminates in an immense fold, the anticlinal, or arch of which, in the case of Switzerland, constitutes the outer crest of the Alps, while the synclinal, or area of depression, has given rise to the great valley of the Po, which appears to be an area of sinking.

The Jura rises gently from the north-west, and culminates in the steep wall which bounds the Central plain of Switzerland.

The Ural Mountains and their continuation, the Islands of Novaya Zemlya, are steep on the eastern side. In fact, the Urals are not so much a chain of mountains, as a tilted surface, with a sudden fracture, and a sunken area to the east.* The Indian Ghats again present a very steep side to the sea. The Himalayas (which in so many respects resemble the Alps), the Rocky Mountains, the Green Mountains, the Alleghanies, etc., are also one-sided; and South America slopes up from the east to the great wall of the Andes which towers over the Pacific Ocean.

The Alps are a most delightful, but most difficult, study, and although we thus get a clue to the general structure of Switzerland, the whole question is extremely complex, and the strata have been crumpled

---

* Suess, *Die Entstehung der Alpen.*

and folded in the most complicated manner, sometimes completely reversed, so that older rocks have been folded back on younger strata, and even in some cases these folds again refolded.

CHAPTER IV.

SNOW AND ICE.—SNOWFIELDS AND GLACIERS.

> "Chaque année je me livre à de nouvelles recherches, et en me procurant un genre de jouissance peu connu du reste des hommes, celui de visiter la nature dans quelques-uns de ses plus hauts sanctuaires, je vais lui demander l'initiation dans quelques-uns de ses mystères, croyant qu'elle n'y admet que ceux qui sacrifient tout pour elle et qui rendent des hommages continuels."—Dolomieu, *Journal des Mines*, 1798.

THE height of the snow-line in the Alps differs according to localities and circumstances, but may be taken as being from 2500 to 2800 metres above the sea-level.

The snowfields are very extensive, the expanse of firn being necessarily greater than that of the glacier proceeding from it.

The annual fall of snow gives rise to a kind of stratification, which however gradually disappears. The action of the wind tends, on the whole, to level the surface, leaving however many gentle undulations, and heaping up the snow in crests and ridges. On the creste of the mountains it often forms cornices, which sometimes project several feet. I shall never forget my sensations, when standing with Tyndall on, as I supposed, the solid summit of the Galenstock, he struck his alpenstock into the snow, and I found that we were only supported on such a cornice projecting over a deep abyss.

When the snow falls at a temperature of $0°$-$12°$, it assumes the form of stars or six-sided crystals.

The region affected by glacial action may be divided into three parts:—

      1. The firn or Névé.
      2. The glacier.
      3. The region of deposit.

THE FIRN OR NÉVÉ.

The snow which falls in the higher Alpine regions, by degrees loses its crystalline form, becomes granular, and is known as Névé or Firn. It can be distinguished at a glance from recent snow by being less brilliantly white, partly because it contains less air, partly because the particles of meteoric and other dust give it a lightly yellowish, grey, or even brownish tinge. Sometimes it is in patches quite red. This is generally due to the presence of a minute alga (Sphærella nivalis). There are, however, several other minute organisms, plants, Infusoria and

Rotifera (Philodina roseola) of a red or brownish colour. The firn is generally firm. When the temperature is low, it becomes quite hard; except on hot days the foot sinks but little into it; usually it remains dry. The water which results from melting sinks into it, and freezes the snow below into a solid mass, which has a more or less stratified appearance, each yearly deposit forming a layer from one to three feet in thickness, which can sometimes be traced even to the lower end of the glacier. The firn attains in many places a great depth. Agassiz estimated that of the Aar glacier at 460 metres.* It moves slowly downwards, and when its upper end terminates against a rock wall, which of course retains its position, a deep gap is formed in spring, known as a Bergschrund, which widens during the summer and autumn, gradually fills up in winter, and reappears the next year.

It is impossible to give any idea in words of the beauty of these high snowfields. The gently curving surfaces, which break with abrupt edges into dark abysses, or sink gently to soft depressions, or meet one another in ridges, the delicate shadows in the curved hollows, the lines of light on the crests, the suggestion of easy movement in the forms, with the

* *Système Glaciaire.*

sensation of complete repose to the eye, the snowy white with an occasional tinge of the most delicate pink, make up a scene of which no picture or photograph can give more than a very inadequate impression, and form an almost irresistible attraction to all true lovers of nature.

The snow would accumulate and increase in thickness indefinitely if it were not removed, (1) by melting and evaporation, (2) by avalanches, and (3) by slow descent into the valley.

AVALANCHES.

Avalanches may be divided into two principal classes; dust avalanches, and ground avalanches.

Dust avalanches generally occur after heavy snowfalls and in still weather, because the snow accumulates on steep slopes until it finally gives way; first in one place and then in another; first slowly, then more rapidly, until at last it rushes down with a noise like thunder.

The falling mass of snow compresses the air, and makes a violent wind, which often does more mischief than the actual avalanche itself. A great part of the snow rests at the foot of the declivity from which it falls, but a part is caught up by the wind and carried to a considerable distance. Such avalanches fall irregularly, as they depend on a variety of circumstances; they cannot therefore be foreseen, and do much damage, often killing even wild animals.

Ground avalanches occur generally in spring, when the snow is thawing. The water runs off under the snow, which thus becomes hollow, only touching the ground in places. A slight shock is sufficient to set it in motion, and it tears away down to the ground, which it leaves exposed. Such avalanches depend therefore on the configuration of the surface, and are in consequence comparatively regular. In many cases they follow the same course year after year. In these tracks, trees cannot grow, but only grass or low bushes.

The front part of the avalanche of course first begins to slacken its speed. The part behind then presses on it, and often pushes over it. Those who have been enveloped in an avalanche all agree, that during the motion they could move with comparative freedom, then at the moment of stopping came extreme pressure, and they found themselves suddenly encased in solid ice. Pressure had caused the particles to freeze suddenly.

Avalanches are often looked on as isolated and exceptional phenomena. This is quite a mistake.

They are an important factor in Alpine life. The amount of snow which they bring down is enormous. Coaz* estimates it in certain districts as equal to 1 metre of snow over the whole district. Without them the higher Alps would be colder, the lower regions hotter and drier. The snow-line would come down lower, many beautiful Alps would be covered with perpetual snow, the glaciers would increase, the climate become more severe, the mountains less habitable. To appreciate the importance of avalanches one must ascend the mountains on a warm day in spring. From every cliff, in every gorge we hear them thundering down all round us. They descend on all sides like hundreds of waterfalls, sometimes in a silver thread, sometimes like a broad cataract. The mountain seems to be shaking off its mantle of snow.

However destructive then they may be at times, avalanches are on the whole a blessing.**

Glaciers.

By the slow action of pressure, and the percolation of water, which freezes as it descends, the firn passes gradually into ice. In cool and snowy

* *Die Lawinen in den Schweizeralpen*, Bern, 1881.
** Heim, *Gletscherkunde*.

summers the thickness of the layer of firn increases. It is deepest in the higher regions, and thins out gradually, until at length ice appears on the actual surface, and the firn passes into a glacier.

Glaciers are in fact rivers of ice, which indeed sometimes widen out into lakes. Glacier ice differs considerably from firn ice, and the molecular process by which the one passes into the other is not yet thoroughly understood. Again, if a piece of ice from a lake is melted in warm air the surface gradually liquefies and the whole remains clear; on the contrary, a piece of compact glacier ice from the deeper part of a glacier if similarly treated behaves very differently; a number of capillary cracks appear, which become more and more evident, and gradually the ice breaks up into irregular, angular, crystalline fragments. These are known as the "grains du glacier" or "Gletscherkorn," and were first described by Hugi.* They increase gradually in size, but how this growth takes place, and whether they are derived from the granules of the firn, is still doubtful. When the firn passes into the glacier they may be about $1/4$ inch across; in the middle part of a large glacier about the size of a walnut, and at the end 4 or even 6 inches in diameter. Those at the end of the Rhone

* *Das Wesen der Gletscher*, 1842.

Glacier vary much in size, but the majority are under an inch across.

In some cases they are tolerably uniform in size, in others large and small are mixed together. On any clean surface of glacier ice they are easily visible, as for instance in the ice tunnel which is so often cut at the end of glaciers. Their surfaces present a series of fine paralled striæ, first noticed by Forel. Glacier ice then may be said to be a granular aggregate of ice crystals. By alternately warming and cooling snow, and saturating it repeatedly with water, Forel found that he produced an ice very similar in structure to that of glaciers. There seems no doubt that this structure considerably facilitates the movements of glaciers.*

Glaciers are generally higher in the middle, and slope down at the two sides owing to the warmth reflected from the rocks. When the valley runs north and south the two sides are equally affected in this respect; but when the direction is east to west or west to east the northern side is most inclined because the rocks lie more in the sun, while those to the south are more in the shade.

* Heim, *Gletscherkunde*.

Movement of Glaciers.

Rendu, afterwards Bishop of Annecy, in 1841 first stated clearly the similarity between the movements of a river and those of a glacier.

Subsequent observations have confirmed Rendu's statements. In fact the glacier may be said really to flow, though of course very slowly.

The movement of a glacier resembles that of a true river, not only generally, but in many details; the centre moves more quickly than the sides; where the course curves, the convex half moves more quickly than the concave, and the surface more quickly than the deeper portions. The movement is more rapid, indeed some three times more quick, in summer than in winter.

The first detailed observations on the movements of glaciers were made independently and almost simultaneously by Agassiz on the Unter-Aär glacier, and by Forbes on the Mer de Glace.

The yearly motion of the Swiss glaciers is estimated at from 50 to 130, or in some exceptional cases even 300 metres. The rapidity differs however considerably, not only in different glaciers, but in different parts of the same glacier; in different years, and different times of year. The remains of Dr.

Hamel's guides, who perished in a crevasse on the Grand Plateau (Mont Blanc) on 20th August 1820, were found in 1861 near the lower end of the Glacier des Boissons, having moved some 4 miles in forty-one years, or nearly at the rate of about 500 feet a year.

It has been calculated that a particle of ice would take at least 250 years to descend from the Strahleck to the lower end of the Under-Aar Glacier; from the summit of the Jungfrau to the end of the Aletsch Glacier about 500 years.

During the Middle Ages the Swiss glaciers appear on the whole to have been increasing in size, and to have reached a maximum about the year 1820. After that they retreated till about 1840, when they again advanced until about 1860, since which time they have greatly diminished, though some are now again commencing to advance. Those of northern Europe appear to be also increasing.* It is, of course, impossible to make any decided forecast as to the future.

Cause of Movement.

But why do glaciers descend?

Scheuchzer in 1705 suggested that the water in

* Heim, *Gletscherkunde*.

the fissures of the glaciers, freezing there and expanding as it froze, was the power which urged them forwards. Altman and Grüner in 1760 endeavoured to explain it by supposing that the glaciers slid over their beds; and no doubt they do so to some extent, but this is quite a subordinate form of movement. Bordier regarded the ice of glaciers "not as a mass entirely rigid and immobile, but as a heap of coagulated matter or as softened wax, flexible and ductile to a certain point." This, the "Viscous" theory, was afterwards most ably advocated by Forbes. No doubt the glacier moves as a viscous body would; but the ice, far from being viscous, is extremely brittle. Crevasses begin as narrow cracks which may be traced for hundreds of yards: a slight difference of inclination of the bed will split the ice from top to bottom. It is, in fact, deficient in that power of extension, which is of the essence of a viscous substance.

The explanation now generally adopted is that which we owe mainly to Tyndall. Faraday in 1850 observed, that when two pieces of thawing ice are placed together they freeze at the point of contact. Most men would have passed over this little observation almost without a thought, or with a mere feeling of temporary surprise. Eminent authorities have differed in the explanation of the fact, but into this part of the question I need not now enter. Sir Joseph Hooker suggested the term "Regelation," by which it is now generally known, and Tyndall has applied it to explain the motion of glaciers.

Place a number of fragments of ice in a basin of water, and they will freeze together wherever they touch. Again, a mass of ice placed in a mould and subjected to pressure breaks in pieces, but as the pieces reunite by regelation they assume the form of the mould, and by a suitable mould the ice may be forced to assume any given form. The Alpine valleys are such moulds. When subject to tension, the ice breaks and crevasses are formed, but under pressure it freezes together again, and thus preserves its continuity.

Professor Helmholtz in his scientific lectures sums up the question in these words—"I do not doubt that Tyndall has assigned the essential and principal cause of glacier-motion, in referring it to fracture and regelation." Other authorities, however, do not regard the problem as being yet by any means solved.* Heim points out that, as in the case of water, a large glacier moves under similar conditions more

* Heim, *Gletscherkunde.*

rapidly than a small one. Many bodies will in small dimensions retain their form, which in larger masses would be unable to support their own weight. A small clay figure will stand where a life-sized model will require support. Sealing-wax breaks under tension like ice, but under even slight pressure gradually modifies its form.

Prof. Heim is convinced that if a mass of lead, corresponding to a glacier, could be placed in a Swiss valley, it would move to a great extent like a glacier. The size of a glacier is therefore an important factor in the question, and throws light on the more rapid movement of the greater glaciers, even when the inclination of the bed is but slight. In Heim's opinion then the weight of the ice is sufficient to account for movement, though the character of the movement and the condition of the glacier is due to fracture and regelation. He sums it up in the statement that gravity is the moving force, and the glacier grains the prevailing mechanical units of movement.

CREVASSES.

The rigidity of ice is well shown by the existence of crevasses. They may be divided into three classes:—

1. Marginal.
2. Transverse.
3. Longitudinal.

The sides of most glaciers are fissured even when the centre is compact. The crevasses do not run in the direction of the glacier, but obliquely to it, enclosing an angle of about 45° (Fig. 29, *m m*) and pointing *upwards,* giving an impression that the centre of the glacier is left behind by the quicker motion of the sides. This was indeed supposed to be the cause, until Agassiz and Forbes proved that, on the contrary, the centre moved most rapidly. Hopkins first showed that the obliquity of the lateral crevasses necessarily followed from the quicker movement of the centre.

Tyndall gives the following illustration:—"Let *A C,* in the annexed figure, be one side of the glacier, and *B D* the other; and let the direction of the motion be that indicated by the arrow. Let *S T* be a transverse slice of the glacier, taken straight across it, say to-day. A few days or weeks hence this slice will have been carried down, and because the centre moves more quickly than the sides it will not remain straight, but will bend into the form *S' T'*.

"Suppose *T i* to be a small square of the original slice near the side of the glacier. In its new position the square will be distorted to the lozenge-shaped figure $T'\,i''$. Fix your attention upon the diagonal $T\,i$ of the square; in the lower position this diagonal, *if the ice could stretch,* would be lengthened to $T'\,i'$. But the ice does not stretch, it breaks, and we have a crevasse formed at right angles to $T'\,i'$. The mere inspection of the diagram will assure you that the crevasse will point obliquely *upward."*

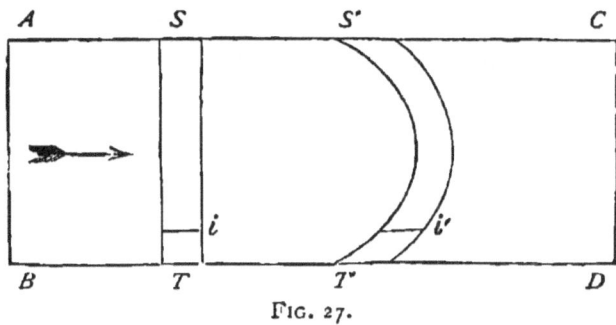

Fig. 27.

Marginal crevasses then arise from the movement of the glacier itself; transverse and longitudinal crevasses are caused by the form of the valley. If the inclination of the bed of a glacier increases, even if the difference be but slight, the ice is strained, and, being incapable of extension, breaks across. Each fresh portion as it passes the brow snaps off in turn, and thus we have a succession of transverse crevasses. In some cases these unite with the transverse fissures, thus forming great curved crevasses, stretching right across the glacier, and of course with the convexity upwards.

Longitudinal crevasses occur wherever a glacier issues from a comparatively narrow defile into a wider plain. The difference of inclination checks its descent; it is pushed from behind, and having room to expand it widens, and in doing so longitudinal crevasses are formed.

The sides of crevasses are of a brilliant blue, and often look as if they were cut out of a mass of beryl. The mountaineers have a traditon that glaciers will tolerate no impurity, and though this is not of course a correct way of stating the question, as a matter of fact the ice is of great purity.

Veined Structure.

Glacier ice very often looks as if it had been carefully and regularly raked. It presents innumerable veins or bands of beautifully blue clear ice, running through the general mass, which is rendered whitish by the presence of innumerable minute air-bubbles. The blue plates are more or less lenticular in structure, sometimes a few inches sometimes many yards in length, but at length gradually fade away.

The whole surface of the glacier in such parts is lined with little grooves and ridges; the more solid blue veins projecting somewhat beyond the whiter ice. This structure is very common, though presenting different degrees of perfection in different glaciers, and different parts of the same glacier. It is rendered the more conspicuous, because the fine particles of dirt are naturally blown, and washed into the furrows. The veins are often oblique, in many cases trans-

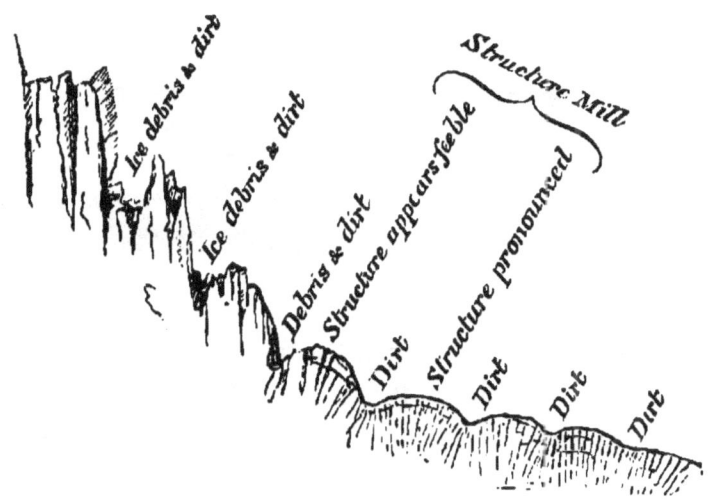

Fig. 28.—Section of Icefall, and Glacier below it, showing origin of Veined Structure.

verse, in some longitudinal, and in others vary in different parts of the glacier.

Here also we owe, I think, the true explanation to Tyndall. We will begin with the oblique veins, which are most marked at the sides, and fade away towards the centre of the glacier. Tyndall points out that if a plastic substance, such as mud, be allowed to flow down a sloping canal, the lateral portions, being held back by the sides, will be outstripped by the centre. Now if three circles (Fig. 29) be stamped on the mud-stream, the central one will retain its form, but the two lateral ones will gradually elongate. The shorter axis in *m m* of each oval is a line of pressure, the longer is a line of strain, consequently

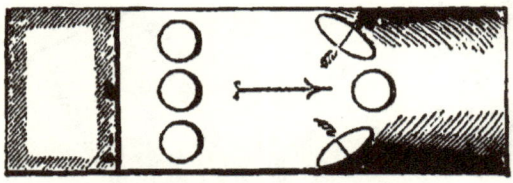

Fig. 29.

along the line *m m,* or across the tension, we have, as already explained, the marginal crevasses; while across the line, or perpendicular to the pressure, we have the veined structure, which is in fact a form of cleavage. Indeed, tension and pressure go together, the one acting at right angles to the other. Passing to the cases of transverse veining, we find if we walk up a glacier presenting this structure that we eventually come to an ice-fall or cascade. At the foot of the fall the ice is compressed, and this gives rise to transverse veining. Longitudinal veining in the same manner arises when two glaciers meet, as for instance the Talèfre and the Léchaud (Fig. 30), where we have transverse pressure and in con-

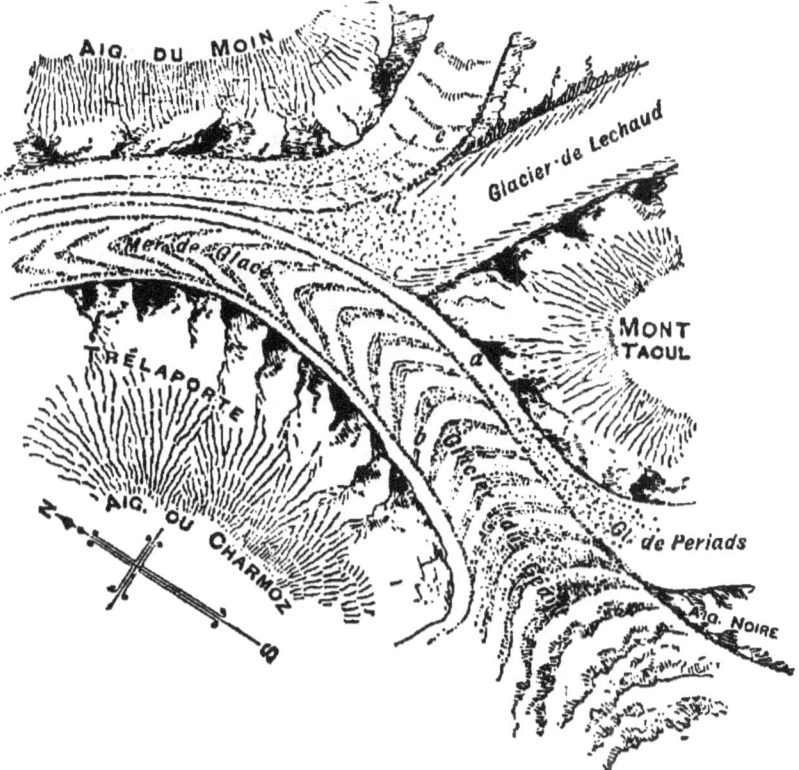

Fig. 30.—Sketch Map of the Mer de Glace.

sequence longitudinal veining. How great must be the pressure in such cases we can faintly realise if we bear in mind that the glaciers which unite to form the great Gorner Glacier have a width of 5200 metres which is compressed to 1000 and further on to 500 metres.

The pressure acts on the ice in two ways— Firstly, in the same manner as it produces lamination in rocks; and secondly, by partially liquefying the ice, thus facilitating the escape of the air-bubbles, which causes its whitish appearance.

Liquid Disks.

The Solar beams also form innumerable liquid disks. As the water occupies less space than the ice each disk is accompanied by a small vacuum, which shines like silver, and is often taken for an air-bubble.

Dirtbands.

If we look down on the Mer de Glace we see (Fig. 30) a series of grey, curved, or bent bands, which follow each other in succession from Trélaporte downwards.

These "dirtbands" have their origin at the ice cascade upon the Glacier du Géant. The glacier is broken at the summit of the ice-fall (Fig. 28), and descends the declivity in a series of transverse ridges. Dust, etc., gradually accumulates in the hollows, and though the ridges are by degrees melted away and finally disappear, the dirt remains, and forms the bands. They are therefore quite superficial. Similar bands occur on other glaciers with ice cascades, and as many as thirty to forty may sometimes be traced.

Moulins.

At night and in winter the glaciers are solemn and silent, but on warm days they are enlivened by innumerable rills of water. Sooner or later these streams reach a crack, down which they rush, and which they gradually form into a deep shaft. These are known as glacier mills or Moulins. Of course the crack moves down with the glacier, but the same cause produces a new crack, so that the process repeats itself over and over again, at approximately the same place. A succession of forsaken Moulins is thus formed. Moulins are often very deep. Desor sounded one on the Finster-Aar glacier which had a depth of 232 metres.

The so-called Giants' caldrons, which will be described further on, are sometimes regarded as indications of ancient glacial action. In the so-called "glacier garden" at Lucerne this no doubt is so; but as a general rule they were probably formed by river action.

In the larger glaciers most of the subglacial rivulets unite under the glacier and flow out at the end in a stream, often under a beautiful blue flat arch generally from 1 to 3 but sometimes even 30 metres in height. In many cases it is possible to enter them for some distance, and galleries are often cut. The ice is a splendid blue, the surface takes a number of gentle curves, and when the light from outside is reflected from the surfaces, it assumes by complementary action a delicate tint of pink.

Moraines.*

The mountain sides which surround glaciers shower down on them fragments, and sometimes immense masses, of rock, which gradually accumulate at the sides and at the end, and are known as "Moraines." When two glaciers meet, a "medial" moraine is formed by the union of two "lateral" moraines (Fig. 30), while the matter carried along under the glacier is known as "ground Moraine." However many glaciers may unite, the moraines keep themselves distinct, and may often be seen for miles stretching up the glacier side by side.

* The word "Moraine" was adopted by Charpentier from the local name used in the Valais, and has now become general.

Even from a distance we may often see by the colour that different moraines, and the two sides of a medial moraine, are composed of different rocks. On the Aar glacier the left half of the medial moraine is composed of dark micaceous Gneiss and Mica Schist; the right half of white Granite. The right lateral moraine of the Puntaiglas glacier, on the south of the Tödi group, is made up of dark greenish Syenite and Granite, the first medial moraine is of titaniferous Syenite, then comes a second of yellowish red Röthidolomite with some Dogger; then several of bluish black Hochgebirgskalk, and lastly the left moraine is of Puntaiglas Granite, and various sedimentary rocks from Verrucano to Eocene.* The Baltora glacier in the Hindu Kush has no less than fifteen moraines of different colours. The different moraines do not mix; and fragments from one side, even of the same moraine, never pass to the other, but move down with the ice, in the same relative positions.

The glacier often rests directly on the solid rock, but in many places there is a layer of clay and stones, to which Ch. Martins gave the name of "ground moraine," and if the underlying rock is examined it will be found to be more or less polished

---

* Heim, *Gletscherkunde*, p. 348.

and striated. The importance of the ground moraine was first pointed out by Martins.* The pressure of the glacier on its bed must be very great. On the Aletsch glacier it has been calculated to be as much as 4 tons to the square decimeter; under the Arctic glaciers it must be much greater. In the winter of 1844 some poles of timber were dropped under the edge of the Aar glacier, in the following year they were found to be crushed to small fragments. Blocks of stone are gradually ground down and reduced to glacial mud. This is so fine that it remains a long time in suspension in water, and gives their milky colour to glacial streams. The ground moraine is no doubt formed in some measure from surface blocks which have found their way through crevasses, and have to a great extent been crushed and reduced to powder; but as ground moraines occur under ice-sheets, such as that of Greenland, when there are scarcely any surface blocks, it is clear that the material is partly derived from the underlying bed.

At the lower end of the glacier a terminal moraine gradually accumulates, which may reach a height of 50, 100, or even 500 metres. They are more or less curved, encircling the lower end of the glacier.

* *Revue des Deux Mondes*, 1847.

The quantity of debris differs greatly in different glaciers: some, as the Rhone, Turtmann, etc., are comparatively free, while others, as the Zinal and the Smutt, have the lower ends almost entirely covered.

It is difficult to give the actual number of glaciers in Switzerland, because some observers would rank as separate glaciers what others would consider as branches, but the number may be taken as between 1500 and 2000. The total area is about 3500 sq. km.

The mean inclination of large glaciers is from 5° to 8°, falling however even to less than a degree. The hanging glaciers are much steeper.

The greatest thickness of the ice can only be estimated. In one place of the Aar glacier Agassiz found a depth of 260 metres without reaching the bottom. From the transfiguration of the surface, however, it may safely be calculated that the ice must attain a thickness in places of 400 or even 500 metres. It has been calculated that the ice of the Görner glacier would be enough to build three Londons.

The distance to which a glacier descends depends partly on the extent of the collecting ground, partly on the configuration of the surface. The Görner glacier advances so far on account of the magnitude of the snow-fields above. In 1818 the lower Grindelwald glacier descended to 983 metres above the sea level. In 1870, it had receded to 1080 metres. The lower limit of the Mer de Glace is 1120 metres. In the Eastern Alps, where the climate is more continental and drier, the general limit is from 1800 to 2300 metres.

Ice Tables.

Small bodies, such as pebbles, dust, insects, etc., tend to sink into the ice. On the other hand larger stones intercept the heat.

On most glaciers may be seen large stones resting on pillars of ice. These are the so-called Ice tables. If the stone be wide and flat, the pillar may reach a considerable height, for the ice immediately under it, being protected from the rays of the sun, melts less rapidly. The tables are rarely horizontal, but lean to the south, that side being more exposed to the sun. Small stones and sand, on the contrary, absorb the heat and melt the snow beneath them, unless indeed there is a sufficient thickness of sand, in which case they intercept the heat and form cones, sometimes ten or even twenty feet in height.

Medial moraines in the same way tend to check

the melting. That on the Aar glacier rises 20, 40, and even 60 metres above the general surface, and from the summit of the Sidelhorn it gives the impression of a wide black wall separating two white rivers. In Greenland such ice-walls have been known to attain a height of 125 metres.

We can hardly have a better introduction to the study of glaciers than a visit to the Rhone glacier. The upper part, which is not shown in the figure, is a magnificent and comparatively smooth ice-field. Then comes a sharp descent, where in a river we should have a cascade or series of cascades, and where the ice breaks into a series of solid waves. The crests gradually melt, and as dust and stones collect in the hollows, and the centre of the glacier moves more rapidly than the sides, we have a succession of dirt bands which curve across the glacier.

Below the fall, the bed of the glacier becomes again comparatively flat; the glacier is squeezed out so as to become considerably wider, and as the ice cannot expand it splits into a number of diverging crevasses. This was much more marked when I first visited the glacier in 1861, and when it was much larger than at present.

If we start from the hotel, after crossing the river,

at a very short distance we come to a bank of loose sand and stones, some angular, some rounded, which curves across the valley, except where it has been washed away by the river. This is the moraine of 1820, and shows the line at which the glacier stood for some years. The Swiss glaciers generally increased till about 1820, then diminished till about 1830, increasing again till about 1860, since which they have retreated considerably. The moraine of 1856, in the case of the Rhone glacier, forms a well-marked ridge some distance within that of 1820.

From that ridge to the foot of the glacier, the valley is occupied by sand and stones in irregular heaps, some of them smoothed and ground by the glacier. This is especially the case with the larger stones, which show a marked difference on their two sides, that turned towards the glacier being smooth, while the lee side is rough and abrupt. Many of the stones were evidently pushed by the glacier along the valley, and have left a furrow behind them. The Rhone wanders more or less over the flat bottom of the valley, and spreads out the material which has been brought down by the glacier.

Here and there on the glacial deposits lie blocks with fresh angles, totally different in appearance from the rounded blocks borne by the glacier. These have been brought down by avalanches.

Near the glacier are two other small moraines, the outer one that of 1885, the inner of 1893. We know that these moraines were deposited by the glacier, and no one who has seen them can doubt that those farther and farther down the valley have had a similar origin.

The Rhone flows from the foot of the glacier in various and varying streams, but especially at one place near the centre of the face of the glacier, where there is a beautiful blue arch, about 25 metres in height.

In 1874 careful measurements were commenced by the Swiss Alpine Club. At first lines of stones were placed annually at the foot of the glacier, but the river washed them away so much that the present limits are laid down annually on a plan. It is found that just as the glacier advances when we have a succession of cold and snowy years, and diminishes when there have been hot and dry periods; so in each year, even when the glacier is on the whole retreating, it advances in two or three of the winter months. Amongst other means of studying the glacier the Commission have placed lines of stones across it at some distances above the fall. One of these lines was arranged in 1874, the stones painted yellow, and their position carefully marked. When they came to the fall they disappeared for four years, after which some of them again emerged at the surface, and some of the central ones have reached the lower end of the glacier, which has retreated some yards from the spot at which they were deposited.

As in many of the most accessible glaciers, a gallery has been cut into the ice, and is well worth a visit.

The exquisite curves into which the ice is melted by the eddying currents of air are very lovely. Again, one can easily trace the glacier grains especially if a little ink or other coloured fluid is rubbed over the surface of the ice, when it runs down between the grains, marking them out with dark lines. Each grain, moreover, shows very fine lines of crystallisation, which are parallel in each grain, but differ in different grains. The chief attraction, however, is naturally the splendid blue colour of the ice, and the lovely pink complementary tints of the reflections from the surface.

CHAPTER V.

ON THE FORMER EXTENSION OF GLACIERS.

> Above me are the Alps
> The palaces of nature, whose vast walls
> Have pinnacled in clouds their snowy scalps,
> And throned Eternity in icy Halls
> Of cold sublimity, where forms and falls
> The avalanche—the thunderbolt of snow!
> All that expands the spirit, yet appals,
> Gather around these summits as to show
> How Earth may pierce to Heaven, yet leave
> vain man below.
> *Childe Harold's Pilgrimage*, BYRON, III. 62.

THE present scenery of Switzerland has been much influenced by the former extension of glaciers, and the fertility of the country is greatly enhanced by the materials which they have brought down from the mountains and spread over the low country. Several of the lakes, moreover, which add so much to its beauty, owe their origin to ancient moraines.

The existence of a glacial period and the great former extension of the Swiss glaciers is proved by four lines of evidence, namely:—

1. Moraines and fluvio-glacial deposits.

THE FORMER EXTENSION OF GLACIERS. 131

2. Erratic blocks.
3. Polished and striated surfaces.
4. Animal and vegetable remains belonging to northern species.

FIG. 31.—View of the Grimsel.

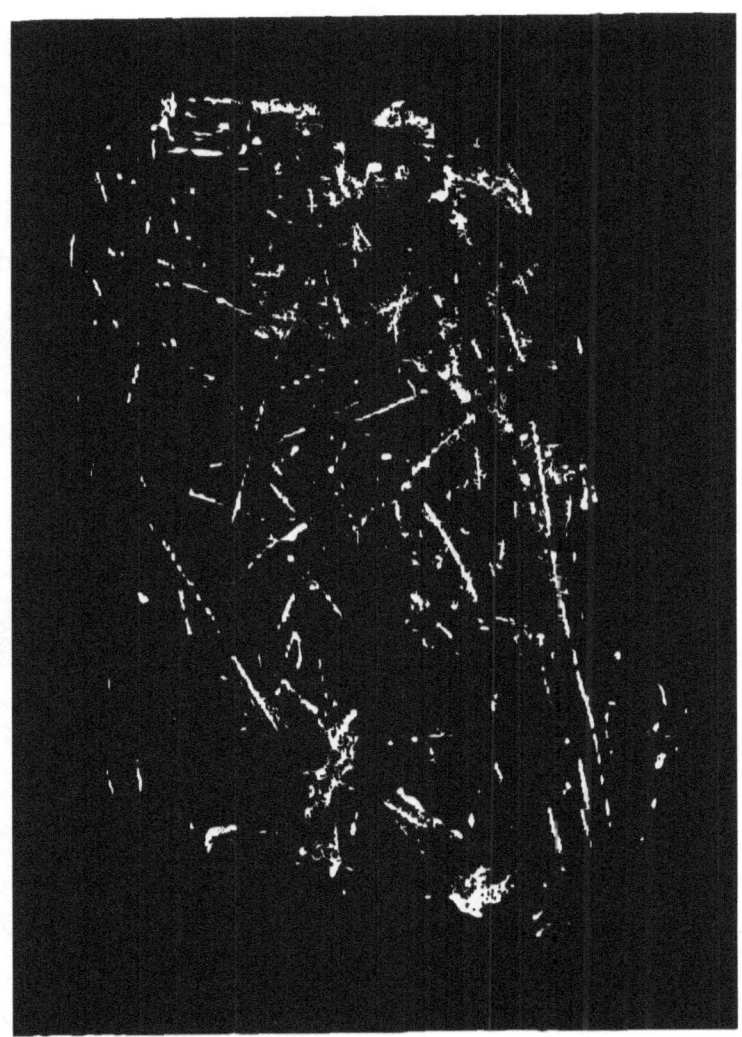

Fig. 32.—Scratched Pebble from the moraine at Zürich.

Glacial Deposits.

Glacial deposits may be classed under two heads:—

1. Moraines.
2. Glacial deposits which have been rearranged by water and may be termed fluvio-glacial.

Moraines are characterised by the presence of polished and striated pebbles, intermixed with more or less angular fragments, often coming from a great distance and yet not rolled, irregularly deposited in sand and mud, which, however, is not stratified.

Fluvio-glacial deposits are composed of the same materials, but more or less rolled, and rearranged by water, like river gravels. They are glacial deposits caught up and carried to a greater or less distance by water.

These two deposits are in intimate relation; they agree in their composition, and differ only as regards stratification. The fluvio-glacial beds, as we come nearer and nearer to their source, are composed of larger and more angular pebbles, while the stratification becomes less and less regular, so that they approximate more and more to the character of true moraine.

The surface of a true glacial deposit is irregular, and presents a succession of hills and valleys, often more or less concentric in outline, and enclosing a central depression (the site of the glacier itself), so that it forms a sort of amphitheatre. See for instance Fig. 33. The Wettingen Feld in the valley of the Limmat is the cone of fluvio-glacial deposits from the ancient moraine of Killwangen.

As glaciers often retreat and then advance again

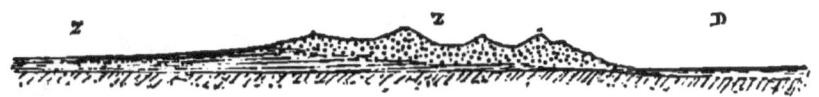

Fig. 33.—Glacial Deposits. *D*, site of ancient glacier; *Z*, moraine; *z*. fluvio-glacial deposits.

the cone of transition in many cases presents alternations of true morainic and fluvio-glacial strata.

When the glacier retreated, the water occupied the central depression between the ice and the moraine, forming a lake. In most cases, however, it cut by degrees through the moraine, and drained the lake. The streams then wandered over the old glacier bed. That the lake naturally overflowed at the lowest point of the moraine, explains why the outflow is often not in the centre of the valley, and occasionally at some distance from the end of the

THE FORMER EXTENSION OF GLACIERS. 135

lake, as for instance, at the Lake of Hallwyl (Fig. 35).

Far down the valleys we find moraines, exactly similar in character to those still being formed along the sides and in front of existing glaciers, and repeated again and again, indicating that glaciers must once have extended far beyond their present areas.

The Rhone glacier occupied the Valais, in which

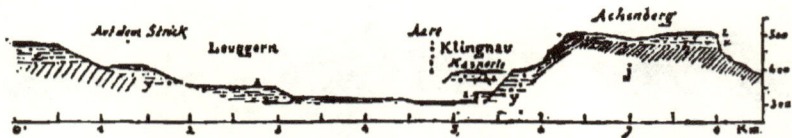

FIG. 34.—Section across the Valley of the Aar above Coblenz. Scale, length 1=100,000; height 1=25,000. *z*, Lower alluvial terrace; *y*, upper alluvium covered by moraines and Löess; *x*, alluvium of the upper plateaux, covered by Löess; Jurassic strata *in situ*.

are several ancient moraines; it filled the whole basin of the Lake of Geneva; and the high terrace of St. Paul above Evian is a moraine, due to the confluence of the ancient glaciers of the Rhone and Dranse; so is also the promontory of Yvoire. Still further down the valley glacial deposits are found along the Rhone as far as, and even beyond Lyons,* and down the Aar to Waldshut.

* Falsan and Chantre, *Les Anciens Glaciers du Bassin du Rhône*. 1880.

Fig. 35.—Map of the country between Lucerne and Aarau.

Fig. 34 represents river terraces and glacial deposits in the valley of the Aar, a short distance above Coblenz.

Passing from the Aar eastwards, in the district of the Wigger, there are important moraines round the Lake of Wauwyl, which was itself the site of a Lake Village carefully studied by Col. Suter, but is now drained.

In the valley of the Suhr is an important terminal moraine at Stafelbach, another at Triengen, while a third encircles and has given origin to the Lake of Sempach.

In the valley of the Winan there is a terminal moraine at Zezwil and another just above Münster.

In the valley of the Aa, are three groups; firstly, one near Schafisheim. Secondly, at the north end of the Lake of Hallwyl are several moraines (Fig. 35); thirdly, between Schafisheim and Egliswyl are three moraines, the inner one encircling a moss, marked Todtenmoos on the map, through which runs the river Aa. Near Nieder Hallwyl is another semicircular moraine enclosing an area of low ground and the end of the lake. It extends along the hill on both sides of the water. The moraines are in parts roughly stratified, and fall away from the lake, having originally sloped no doubt from the great dome of the glacier. A third group encircles the lower end of the Lake of Baldegger.

In the valley of the Reuss is perhaps the finest group of all, consisting of five ridges forming an amphitheatre round the little town of Mellingen. The Heiterberg, between the Reuss and the Limmat, is also encircled by one, which reaches a height of no less than 100 metres.

In the valley of the Limmat there is a fine terminal moraine at Killwangen, another below Schlieren, a third at Zürich, and a fourth forms the bank which crosses the Lake at Rapperschwyl. These terminal moraines are connected by lateral moraines running along the sides of the hills. They do not mark the greatest extension of the glaciers, but indicate places where the glaciers made a stand during their final retreat.

The moraines on the south of the Alps are even more astonishing. Probably from the steeper slope, and more rapid melting under a southern sun, the ends of the glaciers do not appear to have moved so frequently. Hence the terminal moraines are more concentrated, grander, and higher. They form immense amphitheatres terminating in ridges several hundred feet high, and no one seeing them for the first time would for a moment guess their true nature.

The blueness of the sky, moreover, the brilliancy of colouring, the variety and richness of the vegetation, give the moraine scenery of Italy an exquisite beauty with which the north can scarcely vie. Each great valley opening on the plain of Lombardy has its own moraine. At the lower end of the Lago Maggiore at Sesto-Calende are three enormous concentric moraines.* Those of the Lake of Garda are perhaps the largest. They form a series of concentric hills, and attain a height of 300 metres, but those at Ivrea, at the opening of the Val d'Aosta, due to the great glacier procceding from the south flanks of the Mont Blanc range, are the highest and most imposing. They form an amphitheatre round Ivrea. That on the east, known as the "Serra," runs in nearly a straight line from Andrate to Cavaglio, is twenty miles long, and has a height above the valley of 500 metres. The summit line is very uniform. On the outer or eastern side of the great moraine are several other minor ridges. At the right a similar, but less elevated moraine, stretches from Brosso to Strambinello, but it is not so conspicuous, as it rests against the side of the mountain. From Strambinello to Cavaglio it forms a great semicircle which once

* Martins and Gastaldi, "Essai sur les terrains sup. de la Vallée du Po," *Bull. Soc. Géol. de France.* 1850.

probably enclosed a lake, now represented by the Lago di Viverone, Lago di Candia, and some smaller pools. It is nearly bisected by the Dora Baltea. In fact, it is characteristic of the Italian valleys that the surface is comparatively low where the valley debouches into the plain, and then gradually rises towards the Po, forming an amphitheatre whose encircling wall is the outer moraine.*

At several places on the south flank of the Alps, morainic masses are more or less intercalated with younger marine deposits, closely resembling the submarine moraines of the Polar regions, and the Boulder Clay of England and Scotland.

The older moraines are, moreover, less abrupt, and the slopes are more gentle.

Erratic Blocks.

The second class of evidence proving the former extension of glaciers is that presented by erratic blocks, which are often of great size, unrounded, and which have come from a great distance. Many of these are so remarkable that they have struck the imagination of the peasantry, have been attributed to superhuman agency, and have received special names, such as the "Pierres de Niton" in the lake

* Penck, *Vergletscherung der Deutschen Alpen.*

near Geneva, so called from a tradition that in Roman times sacrifices were offered upon them to Neptune. The "Pierre de Crans" near Nyon, is 73 feet long and 20 high.

The "Pierre à Bot," near Neuchâtel, at a height of 2200 feet, is 62 feet in length, 48 in breadth, and 40 feet high. It is of Protogine, and probably came from the Mont Blanc.

Other celebrated erratic blocks are the "Plough-stone," which rises 60 feet above the ground between Erlenbach and Wetzwcil, and contains over 72,000 cubic feet of stone; the Bloc du Trésor near Orsières with a cubic content of 100,000 feet; the Monster block at Montet, near Devent, 160,000; and the largest of all is, I believe, a mass of Serpentine on the Monte Moro, near the Mattmark See, which measures 240,000 cubic feet. These enormous blocks are of course exceptional, but smaller ones are innumerable. In some localities are immense groups — for instance on the hill of Montet, near Devent, at Orsières in the valley of the Dranse D'Entremont above Martigny, at Arpille on the north side of the valley of the Rhone opposite Martigny, and, still further away from the mountains, the entire south slope of the Jura is strewn with Granite blocks. "Between Moliers, Travers, and Fleurier," says De

Luc, "there are as many blocks of primitive rock as if one was in the high Alps."*

One of the most remarkable groups is at Monthey, overlooking the valley of the Rhone below St. Maurice. We have here, says Forbes, "a belt or band of blocks—poised, as it were, on a mountain side, it may be five hundred feet above the alluvial flat through which the Rhone winds below. This belt has no great vertical height, but extends for miles—yes, for miles—along the mountain side, composed of blocks of Granite of thirty, forty, fifty, and sixty feet in the side, not a few, but by hundreds, fantastically balanced on the angles of one another, their grey weather-beaten tops standing out in prominent relief from the verdant slopes of secondary formation on which they rest. For three or four miles there is a path, preserving nearly the same level, leading amidst the gnarled stems of ancient chestnut trees which struggle round and among the pile of blocks, which leaves them barely room to grow: so that numberless combinations of wood and rock are formed where a landscape-painter might spend days in study and enjoyment." **

As already mentioned, these blocks have come

* Agassiz, *Essai sur les Glaciers.*
** Forbes, *Travels through the Alps of Savoy.*

from a great distance. No similar rock occurs in the neighbourhood, and it is often possible to determine the locality from which they have been derived.

For instance, near the Katzensee is a block consisting of a peculiar variety of Granite only known to occur at Ponteljes-Tobel above Trons in the valley of the Rhine. Many blocks of the same rock occur on the right bank of the Lake of Zürich, and they can be followed all the way to their source. Not one occurs to the left of the lake. This could hardly be the case on any other theory than that of transport by a glacier. Again, the "Ploughstone" already mentioned agrees with the fine-grained Melaphyre of the Gandstock in the middle of the Canton of Glarus.

The block of Steinhof near Soleure, which measures 65,000 feet, is probably from the Val de Bagnes.

The Pierre à Bot, as already mentioned, is of Protogine, and has come from the St. Bernard.

It is probable that the ancient glaciers moved more rapidly than their comparatively diminutive descendants of the present day; but at the existing rate of movement the Pierre à Bot would have taken 1000 years to travel from its original home on the chain of Mont Blanc to its present site near Neuchâtel; Whymper calculated that the blocks at Ivrea would have taken a similar period, the Granite blocks of Seeberg would have spent 2000 years and according to Falsan those at Lyons some 4000 years on their long journey.

It is evident that these blocks cannot have been brought by water, both on account of the immense velocity which would have been required to transport such enormous weights, and because, amongst other reasons, their angles are as a rule sharp and unrounded.

Their presence is often attributed to supernatural agency, and many legends grew up round them. Favre* records a remark made to him by a peasant with reference to a great block of Protogine near Sapey. "'Jamais,' disait-ils, 'on a vu une si belle: elle est tout entière, rien de cassé. Et puis, elle est si tranquille. On ne sait pas si les pierres grandissent; mais, il y a 15 ans, je pouvais monter dessus, à présent je ne sais comment cela se fait, mais je n'y puis grimper.'"

Playfair, in 1802, appears to have been the first to compare these erratics with moraines, and to suggest that they were transported by glaciers.

"For the moving of the large masses of rock," he says,** "the most powerful agents without doubt

* *Rech. Géol.*, vol. 1.
** *Illustrations of the Huttonian Theory*, vol. 1.

THE FORMER EXTENSION OF GLACIERS.

which nature employs are the glaciers, those lakes or rivers of ice which are formed in the highest valleys of the Alps, and other mountains of the first order. These great masses are in perpetual motion, together with the innumerable fragments of rock with which they are loaded. These fragments they gradually transport to their utmost boundaries, where a formidable wall ascertains the magnitude, and attests the force, of the great engine by which it was erected." The immense quantity and size of the rocks thus transported have been remarked with astonishment by every observer. Perraudin, a Chamois hunter of the Val de Bagnes, subsequently but independently made the same suggestion to Charpentier. It also occurred to, and was proposed in more detail by Venetz, and at length in 1829 worked out by Charpentier with masterly ability. Agassiz compared the Swiss phenomena with those presented in the north of Europe, and showed that in both cases the country was covered by a sea of ice, from which the highest summits alone emerged.

Charpentier,* and subsequently Guyot,** traced the course of many erratic blocks, and pointed out that as we proceed from the place of origin they

* *Essai sur les Glaciers.*
** *Bull. Soc. Sci. Nat. Neuchâtel,* vol. I.

spread as it were in a fan, and that those from one district do not overlap those from another, as would be the case if they had been distributed by rivers or icebergs: for instance, those of the West Jura come from Mont Blanc and from the Valais, those of the Bernese Jura from the Bernese Oberland, and those of Argovie from the eastern cantons and the Rhine.* Not only are the blocks from each drainage area kept separate, but even, as a rule, those from the two sides of the same valley. I say as a rule, because in some few cases the glaciers appear to have varied in relative dimensions, one encroaching for a time on another, and in its turn being driven back. This however only applies to some few exceptional areas, as for instance between the glaciers of the Linth and the Reuss.

Again, the erratic blocks are specially numerous on the summits and slopes of hills, much more than in valleys: they are not sorted in sizes, but even the largest are found perhaps 50, or even 100, miles from their original site. The smaller blocks are often polished and striated, like those on existing glaciers.

For these and other reasons there can be no

---

* Agassiz, *Etudes sur les Glaciers.*

doubt that they have been carried by glaciers to their present position.

These great blocks, however, imposing as they are, are yet as nothing to the mass of gravel, sand, and mud brought down by the glaciers, carried over intervening ridges and across lakes, and spread over the whole of Switzerland.

The erratic blocks are unfortunately rapidly disappearing, as they are much in demand for building and other purposes. Some of the most remarkable have, however, happily been secured, and will be preserved by the Swiss Scientific Societies.

Considering the immense magnitude of the moraines and the enormous number of erratic blocks, it is evident that the glacial period must have been of very long duration.

Polished and Striated Surfaces.—Roches Moutonnées.

A third class of evidence is that furnished by polished and scratched rock surfaces, which of course are best preserved when the material is hardest. The rocks are sometimes polished like a looking-glass. Such surfaces occur under and round existing glaciers, where there can be no doubt that they are the work of the ice, or rather of the stones contained in it. Fig. 31 is a photograph of the Hospice of the Grimsel, showing a remarkable case of such glaciated rocks. Similar surfaces occur, however, far away from the present glaciers and even in countries where none exist. The grey rounded bosses (Fig. 31) were termed by De Saussure "Roches Moutonnées," from their frizzled surface. The term has been generally adopted, mainly perhaps because at a distance they look not unlike sheep's backs. Smooth rock surfaces may often be seen at the sides of val-

FIG. 36.—Diagram of Crag and Tail.

leys, sometimes at a great height—many hundred or even some thousands of feet above the present river, and far away from the present glaciers, as, for instance, on the slopes of the Jura. They are specially well developed where from a turn in the valley, or any other cause, the ice met with most resistance. The rocks at Martigny are a very fine example.

They do not, however, generally rise to the uppermost ridges, which have therefore (Fig. 37) quite a different character.

De Saussure first noticed the prevalence in the

Alps of smooth, and even polished rock surfaces, but he did not suggest any explanation. Charpentier pointed out that they were due to the action of glaciers. Running water also smooths rocks, but it is almost always easy to distinguish the action of water from that of ice. In the first place, the "Roches Moutonnées" are generally marked by striæ, running in the direction of the valley, and due to

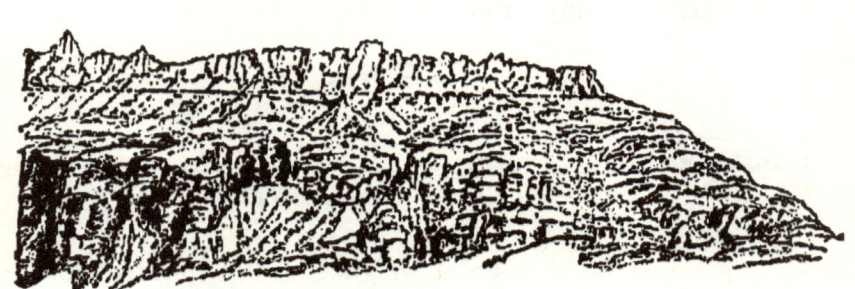

FIG. 37.—View of the Brunberghörner and the Juchlistock, near the Grimsel, showing the upper limit of glacial action.

small stones contained in the ice, and frozen earth. Again, water acts most energetically in the hollows, ice especially on any projecting surface, so that in water-worn surfaces the curves are concave, while on "roches moutonnées" they are convex. The action of water is also much more irregular than that due to ice.

De Saussure was also long ago struck by the fact

that at Chamouni, in the valley of the Aar, and elsewhere, the higher rocks were angular and pointed, while the sides of the valley below were rounded and smooth, but he did not suggest any explanation. Hugi observed the same fact, and attributed it to a difference in the character of the rocks." Desor,* however, in 1841 ascended the Juchliberg, where the contrast is well marked, and satisfied himself that the Granite was absolutely the same. He observed, moreover, that on the smooth Granite, especially on the upper part, were many blocks of Gneiss, brought from the Mieselen and the Ewigschneehorn. These blocks could only have been brought by glaciers, and he concluded, therefore, that the smooth polished surfaces were due to the action of the glacier, and that the rough, angular upper parts were those which had stood above the level of the ice.

Such polished surfaces are by no means confined to the Alpine valleys. Where suitable rocks occur, they are found throughout the central plain and on the Jura, when they are often very well developed, and known locally as Laves. The upper level of the rounded rocks falls with the valley.

On the shores of Norway and Sweden such glaciated surfaces can even be traced under the sea,

* Desor, *Gebirgsbau.*

especially when the water is free from sand. The scratches follow the general direction of the valley, the polished surfaces are on the weather side, and the lee side is the most abrupt, as in Fig. 36. A good example of such smoothed rocks may be seen just in front of the great Hotel at the Maloja.

GIANTS' CALDRONS.

Giants' caldrons are sometimes assumed to be evidence of ancient glacier action. Those at Lucerne doubtless are so, but in other cases similar hollows have been produced by river action.

EVIDENCE DERIVED FROM THE FLORA AND FAUNA.

Another class of evidence is that derived from botany and zoology. Many of the plants now occupying the Swiss mountains are indigenous to the Arctic regions. They could not under existing circumstances cross the intervening plains, but must have occupied them when the climate was colder than it is now, and been driven up into the mountains, like the Marmot and the Chamois, as the temperature rose. The Arctic Willows, the Larch, and Arolla pine, for instance, are Siberian species, and do not occur in Germany.

Here and there also in the drift and the peat-mosses of the lowlands remains have been found of Alpine and Arctic species—the Arolla pine, dwarf birch (Betula nana), Arctic willows (Salix polaris, Salix retusa, and Salix reticulata), Dryas octopetela, Polygonum viviparum, etc.

Moreover, we find living colonies of high Alpine species, the seeds of which can scarcely have been carried by wind, on elevated summits in the lower districts, and in the marshes behind ancient moraines. They cannot have been brought by water, because they occur in some districts not watered by Alpine streams. On the Uetliberg Prof. Heer found two plants which especially characterise moraines—the Alpine toad flax (Linaria Alpina), and a willow-herb (Epilobium Fleischeri). An Alpine fern (Asplenium septentrionale) which is said to be found nowhere else in the Canton of Zürich, occurs on the Plough-stone of Erlenbach. There are two Swiss species of Rhododendron—one with the under surface of the leaves rusty (R. ferrugineum), the other with fringed leaves (R. hirsutum). The latter species prefers a limestone soil and lower regions, so that we should expect to find it prevalent on the Jura. Yet the rusty-leaved species alone occurs there, having probably been brought by the ancient glacier from the

Crystalline mountains of the Simplon and St. Bernard, where it is very abundant.*

The animal kingdom also affords us similar evidence. We find living colonies of Alpine and Arctic animals, especially Insects and Molluscs, on the summits of isolated mountains and in the marshes behind moraines, in association with Alpine plants and erratic blocks. Moreover, just as land animals have retired up the mountains, so have aquatic species been driven into deeper and colder waters—Nephrops norvegicus, for instance, into the depths of the sea at Quarnero, several Arctic animals into the deep waters of the Swedish lakes Wenern and Wettern. In the glacial deposits remains of various Arctic species have been met with. In the gravel-beds near Maidenhead, Charles Kingsley and I found a skull of the Musk Sheep, and remains of the same species, though rare, have since been met with in other parts of Europe. With the Musk Sheep, the Urus, the Aurochs, the Wild Horse, the Mammoth, Hairy Rhinoceros, Reindeer, Elk, the Giant Stag or Irish Elk, Glutton, Ibex, Chamois, Cave Hyæna, Cave Bear, Polar Fox, Lemming, Ptarmigan, Marmot, Snowy Owl, etc., have been also found in glacial

---

* Heer, *Primæval World of Switzerland*, vol. II.

deposits, though fossil remains are rare in the Swiss deposits of this age.

It would be out of place in the present volume to enter into the consideration of the causes which probably led to the existence of the glacial period, or to its probable date. I have in my "*Prehistoric Times*" discussed this question, to which I may refer those who wish to go further into the subject, and I will here only say that I see no sufficient reason to change the opinion (though doubts have recently been thrown on it by Sir H. Howorth and others), that it was mainly due to astronomical causes, and reached its maximum from 50 to 100,000 years ago.

If this explanation be correct it follows that periods of cold and warmth must have followed one another more than once, at intervals of 21,000 years. And in accordance with this we find, as Morlot long ago pointed out, that the glaciers have advanced and retreated more than once.

Beds indicating warmer conditions are interposed between glacial deposits, and the Swiss and South German geologists believe that there were three periods of cold with milder intervals. In Scotland James Geikie and others have brought forward evidence of more numerous oscillations.*

* *The Great Ice Age.*

Morlot was primarily led to this conclusion by his observations in the valley of the Dranse, south of Thonon on the Lake of Geneva, which I had the pleasure of visiting under his guidance. In this gorge between two well-marked glacial deposits is a deposit indicating a milder climate.

Again, at several places in the Canton of Zürich are beds of lignite, sufficiently thick to have been worked for fuel. They are intercalated between glacial deposits; they indicate a luxuriant vegetation and consequently a mild climate; they contain, moreover, remains of animals, such as the Hippopotamus, which could not support great cold. This can only be accounted for, I think, by assuming that these groups of animals occupied the country alternately.

Moraines which have been long exposed to the atmosphere become gradually modified at the surface. The pebbles are much weathered and sometimes quite disintegrated, even those of Granite crumbling into a sort of clay while retaining their original form. The layer affected may have a thickness of one to two feet or even more. This weathered crust often assumes a reddish colour, whence it is called by Italian geologists "Ferretto."

Where an old moraine has after a long interval been covered by a later one, the Ferretto enables us to distinguish between the two. It is in fact a strong confirmation of the existence of inter-glacial periods, during which the glaciers retreated, and a more genial climate prevailed. At Ivrea, for instance, the presence of Ferretto shows that the gigantic moraine known as the Serra was not formed during one long continuous glaciation. The moraines which are coated with Ferretto occupy as a rule the outer side of the morainic amphitheatre, and are covered on their inner edges by the later and inner moraines. Lignite beds also occur on the south of the Alps.* One of the places where an inter-glacial period is most clearly shown is in the valley of the Inn. At Hottingen, close to Innsbruck, is a great fluvio-glacial deposit, reposing on a ground moraine at a height of 1300 metres above the bottom of the valley, and capped to a height of 1900 metres by another. In these fluvio-glacial beds forty-one species of plants have been found and studied by M. Wettstein. Of these twenty-nine now live in the immediate neighbourhood, six in the Tyrol, but at a lower level, six further south, and four have not been determined. Here then we have evidence that the valley of the Inn was (firstly) filled by a glacier to the height of 1300

* Rütimeyer, *Über Pliocene und Eisperiode auf beiden Seiten der Alpen.*

metres, (secondly) that then followed a period with a climate somewhat milder than the present, succeeded (thirdly) by another glacial period, during which the valley was again filled by ice to a depth of 1900 metres.*

The first Age is represented by ground moraine, and by "Deckenschotter"; a diluvial gravel, curiously characterised by the presence of rounded hollows. These were formerly occupied by pebbles, which have been dissolved and washed away through the hard but permeable matrix.

With the exception of one or two heights, as for instance the Napf, there is, on the whole of the Central Plain between the Jura and the Rhine, no considerable area where traces of former glacial action are not to be met with. They attain in places a great thickness, sometimes even more than 400 metres.

It seems at first therefore remarkable that no terminal moraines are known which can be referred to this period. But it must be remembered that the whole country was covered by ice, with the exception of the very highest parts. Hence no doubt, as is the case in Greenland now, the surface of the ice was very free from debris, and hence, perhaps, the

* Penck, *Vergletscherung der Deutschen Alpen.*

peripheral glacial deposits are only represented by ground moraine.

The second Ice Age is represented by the moraines high up on the hills overlooking the valleys; and the third by moraines which form more or less complete ridges curving across the valleys, and along the slopes. It is possible that the glaciers may in some cases have been pushed forwards again over the inner moraines. At Hallwyl, for instance, the moraine immediately encircling the lake is very flat, which Dr. Mühlberg thinks may be thus accounted for.

LIMITS OF THE ANCIENT GLACIERS.

The evidence seems then conclusive that the glaciers were once far larger than at present, and the facts already summarised give some indication of the extent. Beginning with the Rhone Glacier, the former upper limit of the ice at Oberwald was 2766 metres, or 1400 above the river;* at Viesch it was 2700, or 1700 above the river; at Leuk 2100, or 1470 above the river; at Martigny 2080, or 1620; at Geneva 1300, or 950 metres above the Lake.** On the slopes of the Jura it rises highest at Chasseron, north-west of Neuchâtel, opposite the valley of the

* Falsan and Chantre, *Anc. Glaciers du V. du Rhone*, vol. II.
** Favre, *Description Géol. du Canton de Genève*, vol. I.

Rhone, where it attains an elevation of over 1350 metres, or 977 above the lake, descending gradually to the plain on one side at Soleure, on the other at Gex. At Neuchâtel, the erratic blocks form a band about 800 feet above the lake. Above and below that line they rapidly diminish in number.

The Rhone glacier then, at the period of its greatest extension,* not only occupied the whole Valais and the Lake of Geneva, but rising on the Jura to a height of 1350 metres, crossed the Vuache, descended into the present Rhone valley, sweeping round by Bourg, Trévoux, Lyons, and Vienne on one side, sent a wing beyond Pontarlier as far as Salins and Ornans, and extended down the valley of the Aar as far as Waldshut, almost meeting the western extremity of the glacier of the Rhine.

The ancient glacier of the Rhine occupied the Lake of Walen, the whole valley of the Thur as far as Schaffhausen, the Klettgau, and almost to Waldshut, filled up the Lake of Constance, extending considerably to the north down the Danube as far as Sigmaringen, while for some distance its northern end followed the present watershed between the regions of the Rhine and the Danube.

Thus the two great glaciers of the Rhone and

* See Favre, *Carte des Anciens Glaciers de la Suisse.*

the Rhine almost enclosed those of the Aar, the Reuss, and the Limmat. That of the Aar extended as far as Berne, where there is a very fine moraine.

The glacier of the Reuss extended to Aarau, and down the Valley of the Aar to Coblenz. On the east it filled the Lakes of Egeri and Zug, extended along the Albis to the Uetliberg, and to Schlieren on the Limmat, following the valley down to Coblenz.

The glacier of the Limmat was bounded on the west by that of the Reuss; on the east from Wesen on the Lake of Walen, to the Rhine at Eglisau, following the valley to Coblenz, where therefore these four great glaciers met.

The glaciers of the Mont Blanc range not only filled the Valley of Chamouni and the country to the west as far as, and beyond, the Lake of Bourget, but flowed over to the east and joined that of the Rhone.

In fact a sea of ice covered the whole country, with the exception of some mountain tops, from Lyons to Basle, along the Rhine and the Lake of Constance across Bavaria, extending to Munich, and beyond Salzburg.

The extension of the glaciers does not however necessarily imply any very extreme climate.

Paradoxical as it may appear, glaciers require heat as well as cold: heat to create the vapour, which again condenses as snow. A succession of damp summers would do more to enlarge the glaciers than a series of cold seasons. Leblanc* estimated that the glacial period need not have had an average temperature of more than 7 degrees centigrade below the present, and other great authorities consider that at anyrate a fall of even 5° would suffice.

The temperature decreases 1° for about every 188 metres. A fall of 5° would = 940 metres. The present snow-line being 2700 metres, would descend to 1760 metres, and the lower limit of the glaciers from 1200 metres to 360 or somewhat below Geneva, the level of which is 375. It would indeed be even lower, because the greater the snow-field, the further the glacier descends.

We have no evidence of the existence of Man in pre-glacial times, and whether he inhabited Switzerland during the inter-glacial period is still uncertain. Rütimeyer has described certain pieces of wood belonging to that period, which have been cut by some sharp instrument, and which are so arranged as to form a sort of basket-work. They certainly appear

* *Bull. Soc. Geol. France.* 1843.

to be due to human workmanship, but the evidence is not altogether conclusive.

It has happened no doubt to many of us to stand on some mountain-top when the surrounding summits have been covered with snow, and the intervening valleys have been filled with a thick white mist, which, especially in the early morning light, can hardly be distinguished from snow. In such a case, we have before us a scene closely resembling that which the country must have presented while it was enveloped by the ice of the glacial period.

The geologists of Bavaria have brought forward strong evidence for the belief that in Bavaria and Swabia there were three periods of great extension of the glaciers with intervals of a milder climate; and Dr. Du Pasquier, who has especially studied the fluvio-glacial deposits of Switzerland, considers that they confirm this view.

The first cold period is, he considers, represented by the so-called "Deckenschotter," of which perhaps the best known example is that on the summit of the Uetliberg near Zürich, at a height of 400 metres above the lake. It is a coarse gravel, more or less cemented together, and in which many of the pebbles have perished and disappeared, leaving rounded cavities.* This deposit originally formed a more or less continuous sheet, from 30 to 50 metres in thickness, deposited by the water flowing from the melting glaciers, but has been to a great extent removed, fragments only remaining here and there on the high ground. It is remarkable that it contains no traces of Julier or Puntaiglas Granite,** probably because these rocks were still covered by the Crystalline schists. The lateral Moraines of this period are unknown, but the ground Moraines are sometimes well developed. Under the Deckenschotter on the Uetliberg they attain a thickness of 2 to 20 metres. They were probably for the most part destroyed by the glaciers during the Second Ice Age.

The Second Ice Age is represented by gravel-beds, still far above the present valleys, though at a lower level, and by outer and upper moraines, for an instance in the Zürich district those of Höngg, of the Albis, etc. The terminal moraines of this period were however probably beyond the boundaries of Switzerland.

The Third Ice Age is indicated by the lower terraces and the moraines in the valleys. In that of Zürich, the Moraine of Killwangen was pro-

---

* This structure does not occur in the true Nagelflue.
** Du Pasquier, *Beitr. z. Geol. K. d. Schw.*, L. XXXI.

bably the outermost, while those of Zürich and Rapperschwyl represented long periods of arrest and standstill of the glaciers during their general retreat.

In theory this explanation is clear and simple, but it is not always easy to identify the beds. The "Deckenschotter," or upper and older bed, can indeed be generally recognised by the numerous cavities, the "rotten" condition of many of the pebbles, by its being much more frequently cemented together, and in some districts by the nature of the pebbles; in the Zürich Valley, for instance, by the absence or great scarcity of Sernifite and of the Alpine siliceous rocks, and by the frequency of Hochgebirgskalk, which does not occur in the Miocene Nagelflue;* but there are many glacial deposits the exact age of which is very uncertain.

The following table gives the periods, the deposits, and the great characteristic Mammalia, according to Dr. Du Pasquier:** —

* Appeli, *Beitr. z. Geol. K. d. Schw.*, L. XXXIV.
** *Beitr. z. Geol. K. d. Schw.*, L. XXXI.

TABLE OF FLUVIO-GLACIAL DEPOSITS.

CLIMATIC CONDITIONS.	GEOLOGICAL PERIOD.	DEPOSITS.	ORGANIC REMAINS.	CORRESPONDING DEPOSITS IN ENGLAND.
Last Glacial Period.	Upper Pleistocene.	Inner Moraines. Lower Terrace grands.	Mammoth (Elephas primigenius). Rhinoceros tichorhinus.	River Gravels.
Short Inter-glacial Period.	Middle Pleistocene.	Loess Beds of Coal.	Elephas antiquus. Rhinoceros Merckii Hippopotamus major.	...
Middle Glacial Period.	...	Outer Moraines. High Terrace gravels.	...	...
Long Inter-glacial Period.	Lower Pleistocene.		Elephas meridionalis.	Forest Bed.
First Glacial Period.	Upper Pliocene.	"Deckenschotter."	Elephas meridionalis. Mastodon arvernensis.	Norwich Crag.

The Glacial periods were in general, in Dr. Du asquier's opinion, so far as the central Swiss valleys were concerned, periods of deposit, the inter-glacial, periods of excavation.

CHAPTER VI.

VALLEYS.

VALLEYS and rivers are so closely associated with one another, that we generally think of them as inseparably connected; and indeed there are but few valleys which have not been deepened and profoundly modified by the action of water.

Nevertheless many valleys are "tectonic," that is to say, they are due, or stand in a definite relation, to geological structure; and there are some details of valley modelling, which are independent of water action, and which it may be convenient to consider separately.

As already mentioned the plain of Lombardy is a valley of subsidence, the lower limb, as it were, of the great arch of the Alps. It has not been excavated by the Po; on the contrary, that river has been for ages occupied in filling it up, and at Milan a boring was sunk 162 metres without reaching the bottom of the river deposits.*

* Penck, *Morphologie der Erde*, vol. II.

The valley of the Rhine below Basle is also a line of subsidence, and the two Crystalline regions of the Black Forest and the Vosges were once continuous.

Valleys belong to several different classes, and in Switzerland have received special names, such as Vals, Combes, Cluses (Clausa, closed), Ruz, Cirques, etc., which, however, do not cover all the different kinds, and are not always used in the same sense.

In many cases valleys follow the "strike" or direction of the strata, in which case they are termed, as first suggested by De Saussure, longitudinal valleys; while in others they cut across the strata and are known as transverse or cross valleys, or cluses.

Longitudinal valleys again, as Escher von der Linth first pointed out, are of three distinct kinds.

Synclinal valleys (see *ante*, p. 69) occupy the depressions of folded strata. Many of the Jura valleys belong to this class. They are generally broad.

Anticlinal valleys are those which arise when the arch between two synclinals is broken, and the action of water being thus facilitated, a valley is formed, as for instance the Justithal (Fig. 126, vol. II. p. 164), which opens on the Lake of Thun.

In both these classes the strata are the same on the two sides of the valley. A third class of longitudinal valleys is due to the outcrop of layers of different hardness.

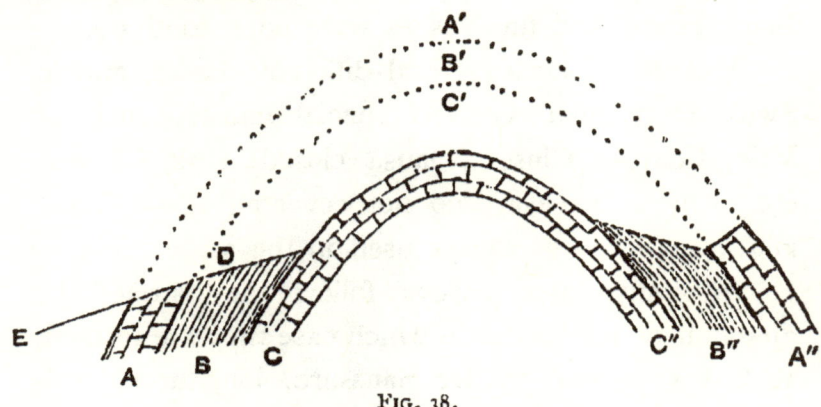

Fig. 38.

In such cases the strata on the two sides are dissimilar; such valleys are known in Switzerland as "Combes."*

Suppose a fractured anticlinal ($AA'A''$, Fig. 38) has been lowered by denudation to $AC'A''$, and is drained by a stream running from C' to E. If the strata are of different degrees of hardness, a soft stratum $BB'B''$ be-

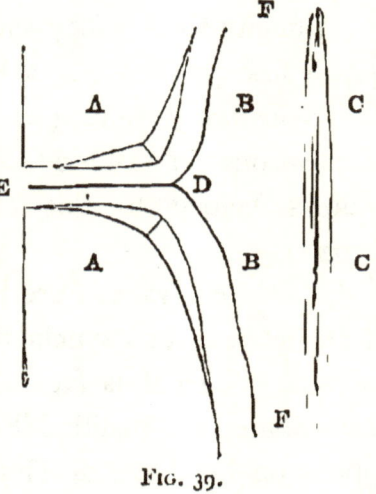

Fig. 39.

* In this country the word "Combe" is often used as synonymous with "cirque."

tween two harder ones A and C will here and there be brought to the surface.

In such a case, owing to the greater softness of the stratum B, secondary streams will often cut their way back as in Fig. 39, FF, thus forming longitudinal valleys parallel to the ridge, the sides being formed by the harder strata AC. Such valleys (Fig. 40) are common in the Jura.

Sometimes there may be two or even three such "Combes" along a main valley, as for instance (Fig. 41) between Mont Tendre and the valley of the Orbe, where we have four ridges of harder strata, Urgonian, Neocomian, Valangian, and finally Portland rock enclos-

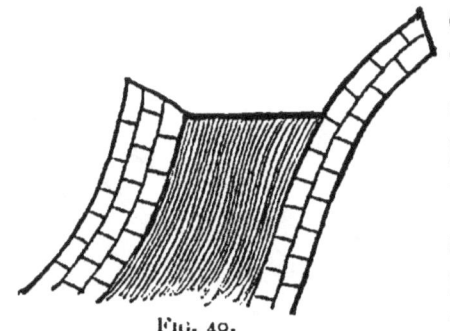

Fig. 40.

ing three combes due to the existence of softer layers.

It is obvious that in this case the transverse valley DE (Fig. 39) is older than the longitudinal valley FF.

A glance at any geological map of Switzerland will show that many rivers run along the boundary, that is at the outcrop, of strata.

The long lines of escarpment which stretch for

miles across country, and were long supposed to be ancient coast lines, are now ascertained, mainly by the researches of Whitaker, to be due to the differential action of subaerial causes. The Chalk escarpments in our own country and the great wall of the Bernese Oberland are of this character. That the longitudinal valleys owe their origin to the same cause as the mountain chains, may safely be inferred from the fact that they follow the same direction. They are in fact negative mountain chains.

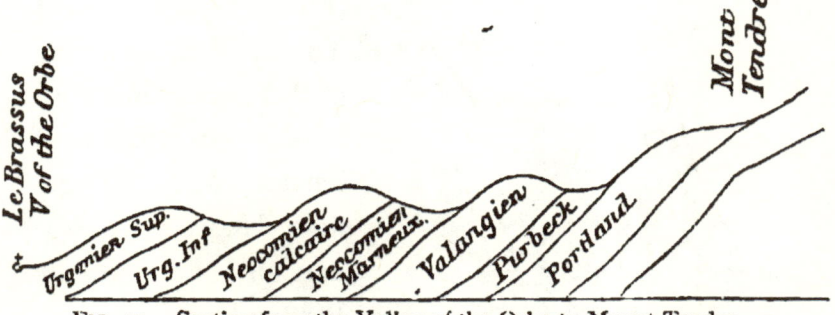

Fig. 41.—Section from the Valley of the Orbe to Mount Tendre.

Transverse Valleys.

Transverse valleys cross the strata more or less at right angles. They are generally narrow, and often form deep gorges, more or less encumbered by fallen rock, and the harder the rock the narrower the valley.

Their character is greatly influenced by the

nature of the strata, their inclination, and whether the fall coincides with, or is in opposition to, that of the beds.

Unless, however, the fall of the ground coincides exactly with that of the strata, a river running along a transverse valley will generally cross here and there harder layers which give rise to cataracts or waterfalls.

When the strata are horizontal the action of running water is comparatively slow. Steeply inclined or vertical strata on the other hand greatly facilitate erosion. Not only does the force of gravity take part in the labour, but the water sinks in more easily, and both chemical and mechanical disintegration is thus much increased.

Hence it is that while cross valleys often drain longitudinal valleys, the reverse seldom happens. Cross valleys in fact dominate longitudinal valleys.

Another respect in which, so far as Switzerland is concerned, the longitudinal differ from the transverse valleys, is that the former run approximately east to west, the latter north to south. This makes a great difference in their general aspect. In the transverse valleys not only do the two sides consist of similar rocks, but both receive approximately the same amount of light and sunshine, so that the vegetation grows under more or less similar conditions.

In the longitudinal valleys, on the contrary, not only are the strata often different on the two sides, but the northern side, which looks to the south, receives more sun, while the southern side is more in shadow. The contrast is strongly shown in the Valais itself, where the south side is green and well wooded, the north, on the contrary, comparatively dry and bare.

In some places, for instance in the valley of the Rhone below Visp, the green lines of vegetation which follow the "Bisses" or artificial water-courses are very conspicuous.

On the Lake of Zürich, though the vegetation is the same on both sides—woods and meadows and vineyards—the distribution is quite different. Both sides of the Lake are terraced, so that we have flat zones and steep slopes. On the northeast side the slopes get more sun, and hence the vines are planted on them, while the meadows and woods are on the terraces. On the west, however, the terraces get more sun, and consequently the vines are on the terraces and the meadows and woods on the slopes.

There is another class of valleys, namely, those which are due to lines of fracture or dislocation, and which may be termed fault-valleys. They are, however, comparatively rare.

One and the same river may be of a very different character in different parts of its course. It may run at one place in a longitudinal, at another in a transverse valley. The Rhone for instance occupies a transverse valley from the glacier nearly to Oberwald, a longitudinal valley from Oberwald to Martigny, and a cross valley from Martigny to the lake.

If we look at an ordinary map of Switzerland, we can at first sight trace but little connection between the river courses and the mountain chains. If, however, the map is coloured geologically, we see at once that the strata run approximately from S.W. to N.E. and that the rivers fall into two groups running either in the same line or in one at right angles to it.

The central mountains are mainly composed of Gneiss, Granite, and Crystalline Schists; the line of junction between these rocks and the Secondary and Tertiary strata on the north, runs, speaking roughly, from Hyères to Grenoble, and then by Albertville, Sion, Chur, Innsbruck, Radstadt, and Hieflau, towards Vienna. This line is followed (in some parts of their course) by the Isère, the Rhone, the Reuss,

VALLEYS.

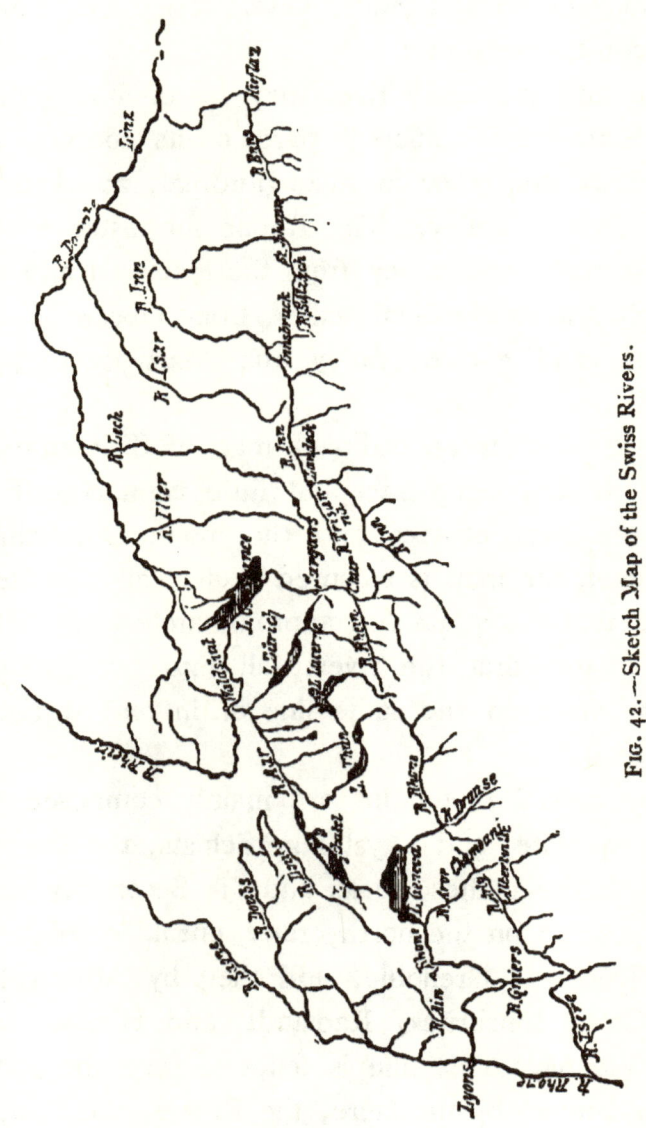

Fig. 42.—Sketch Map of the Swiss Rivers.

the Rhine, the Inn, and the Enns. One of the great folds shortly described in the preceding chapter runs up the Isère, along the Chamouni Valley, up the Rhone, through the Urseren Thal, down the Rhine Valley to Chur, along the Inn nearly to Kufstein, and for some distance along the Enns. Thus, then, five great rivers have taken advantage of this main fold, each of them eventually breaking through into a transverse valley. The origin of the valley is therefore not due to the rivers running through it. Again, as a glance at the foregoing map (Fig. 42) shows, the great valley of the Aar from the Lake of Neuchâtel to Coblenz is continued in the course of the Upper Danube.

The Pusterthal in the Tyrol offers us an interesting case of what is obviously a single valley, slightly raised, however, in the centre, near Toblach, so that from this point the water flows in opposite directions —the Drau eastward, and the Rienz westward. In this case the elevation is single and slight: in the main valley of Switzerland there are several watersheds, and they are much loftier, still we may, I think, regard that of the Arve (see Fig. 42) from Les Houches to the Col de Balme, of the Rhone from Martigny to its source, of the Urseren Thal, of the Vorder Rhine from its source to Chur, of the Inn from Landeck to below Innsbruck, even perhaps of the Enns from Radstadt to Hieflau as in one sense a single valley, due to one of these longitudinal folds, but interrupted by bosses of Gneiss and Granite—one culminating in Mont Blanc, and another in the St. Gotthard—which have separated the waters of the Isère, the Rhone, the Vorder Rhine, the Inn, and the Enns. That the Valley of Chamouni, the Valais, the Urseren Thal and the Vorder Rhine really form part of one great fold is further shown by the presence of a belt of Jurassic strata nipped in, as it were, between the Crystalline rocks.

This great valley then, though immensely deepened and widened by erosion, cannot owe its origin or direction to river action, because it is occupied in different parts by different rivers running in opposite directions. We have in fact one great valley, but several rivers. It is therefore due to one original cause; it is, to use a technical term "geotectonic," and is due to the great lateral compression from S.E. to N.W. which has thrown Switzerland into a succession of great folds.

A similar case is that of the Val Ferret. The depth is no doubt mainly due to erosion, but it follows the tract of Jurassic strata which lie at the foot of the great mountain wall of the Mont Blanc range. No one who looks at the map can for an instant doubt that it is in reality a single valley; but it falls into three parts—the eastern portion is occupied by a branch of the Dranse running to the N. E.; the centre by the Doire running S.W., and the west by another branch of the Doire running N.E., the two, meeting at the foot of the Glacier de la Brenva, fall into a transverse valley and run S.E. towards Courmayeur and Aosta. Again the great valley which has given rise to the Lakes of Neuchâtel and Bienne, and which follows the course of the Aar from Soleure to Brugg, reappears in the course of the Danube below Donaueschingen.

In some respects the courses of the rivers indicate the original configuration of the surface even better than the mountains.

Many rivers after running for some distance along the strike (see p. 66) of the strata, change their direction, not turning in a grand curve, but suddenly breaking away at right angles, as for instance the Rhone at Martigny, the Aar near Brugg, the Rhine near Chur, the Inn near Kufstein.

But why should the rivers, after running for a certain distance in the direction of the main axis, so often break away into cross valleys? The explanation usually given is that transverse streams have cut their way back, and thus tapped the valley. This is no doubt true in some cases, but cannot be accepted, I think, as a general explanation.

Prof. Bonney* called attention to this tendency in his second lecture on the "Growth and Sculpture of the Alps." "On considering," he says, "the general disposition of the rocks constituting the Alpine chain, we perceive that, in addition to the long curving folds which determine the general direction of the component ranges, they give indications of a cross folding. The axis of these minor undulations run from about N.N.E. to S.S.W."

He suggests three possible explanations:—(1) "That the Alps are the consequence of a series of independent movements, not simultaneous, so that the chain results from the accretion laterally of an independent series of wave-like uplifts; (2) that the chain was defined in its general outline by a series of thrusts proceeding outward from the basin of the North Italian plain, and afterwards folded transversely by a new set of thrusts acting at right angles to a N.N.E. line; (3) that the transverse disturbances are the older, and that the floor on which the Secondary deposits were laid down had already been disposed

* *Alpine Journal*, Nov. 1888.

in parallel folds, trending roughly in the above direction."

He adopts the third hypothesis. He considers that the transverse wrinkles were perhaps Triassic, "not improbably post-Carboniferous," and therefore far older than the main longitudinal folds. "Still," he continues, "though I incline to this view, the question is so complicated that I do not feel justified in expressing a strong opinion, and rather throw out the idea for consideration than press it for acceptance. All that I will say is that I find it impossible to explain the existing structure of the Alps by a single connected series of earth movements."

Under these circumstances I have ventured* to make the following suggestion. If the elevation of the Swiss mountains be due to cooling and contraction leading to subsidence as suggested in page 58, it is evident, though, so far as I am aware, this has not hitherto been pointed out, that, as already suggested, the compression and consequent folding of the strata (Fig. 43) would not be in the direction of $A B$ only, but also at right angles to it, in the direction $A C$, though the amount of folding might be much greater in one direction than in the other. Thus in the case of Switzerland, as the main folds

* *Beauties of Nature.*

run S.W. and N.E. the subsidiary ones would be N.W. and S.E.

If these considerations are correct it follows that, though the main valleys of Switzerland have been

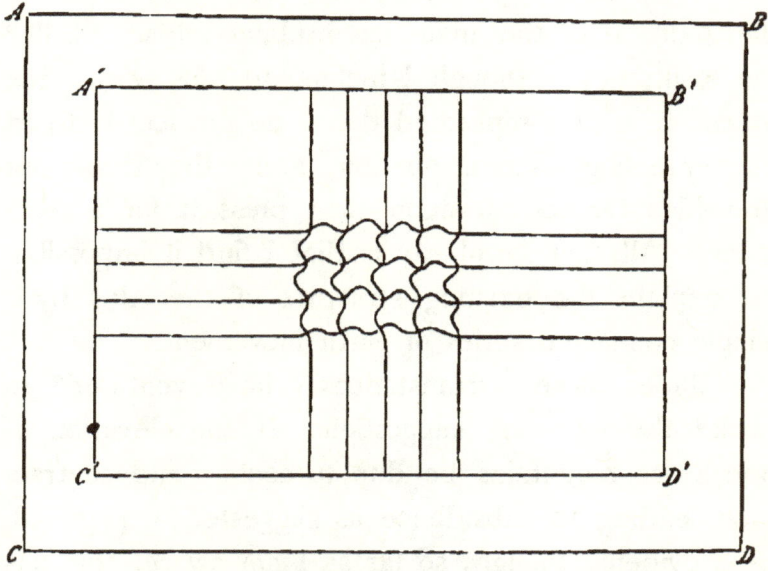

FIG. 43.—Diagram in illustration of Mountain Structure.

immensely deepened and widened by rivers, their original course was determined by tectonic causes.

Again, they indicate why the long valleys are not more continuous. If we look for instance at a map of the Jura, we see that though the ridges follow the same general curve from one end to the other, they are not continuous, but form a succession of similar, but detached ridges. Moreover even when a valley is continuous for many miles it is interrupted here and there by the cross folds.

These considerations then seem to account for the two main directions of the Swiss valleys.

I must add, however, that in Prof. Heim's opinion the cross folds occur in other parts of the Earth's surface; and such bosses as the Furca and the Ober Alp are merely the battle-grounds of different river systems, the lower levels being due to more rapid denudation.

Cirques.

In some cases valleys end in a steep amphitheatre known as a "Cirque."

Cirques are characteristic of calcareous districts. They occur especially where a permeable bed rests on an impervious substratum. Under such circumstances a spring, in many cases intermittent, issues at the junction and gradually eats back into the upper stratum, forming at first a semicircular enclave, which becomes gradually elliptic, and as time passes on more and more elongated, but always with a steep terminal slope. In the Jura, cirques are numerous, and in many cases a marly bed supplies the impermeable stratum.

The Creux du Vent, and the Cirque de St. Sulpice are two of the finest examples.

Terraces.

As regards the sides of valleys, other things being equal, the harder the rocks the steeper will the slope of the sides be. Very hard rocks indeed are often almost, or for some distances quite, perpendicular. The slope may be uniform in cases where the strata are similar and of great thickness, as for instance in the valley of the Reuss above Amsteg where the Bristenstock forms a grand pyramid of Crystalline rock, or where the slope coincides with the dip of the strata, as in the valley of Lauterbrunnen, where the right side of the valley presents immense sheets of Jurassic rock.

In most cases, however, some of the strata along the side of the valley are harder than others, and the consequence is that we have a succession of terraces; gentler slopes indicating the softer, and steeper ones the harder beds.

Figs. 44 and 45 show some terraces in the valley of the Bienne (Jura) due to the presence of hard calcareous layers.

These "weather" terraces (Figs. 54, 55 pp. 202, 203) must not be confused with the "river terraces" which will be described in the next chapter. River terraces have no relation to the rock and follow the slope of the river, while weather terraces follow the lines of the strata.

This consideration throws light on the cases in

Fig. 44.—Weather Terraces in the Valley of Bienne (Jura).

which a river valley expands and contracts, perhaps several times in succession.

We often, as we ascend a river, after passing along a comparatively flat plain, find ourselves in a narrow defile, down which the water rushes in an impetuous torrent, but at the summit of which, to our surprise, we find another broad flat expanse. This is especially the case with rivers running in a transverse valley, that is to say of a valley lying at

Fig. 45.—Section showing Weather Terraces.

right angles to the "strike" or direction of the strata (such, for instance, as the Reuss), the water acts more effectively than in cross rocks which in many cases differ in hardness, and which therefore of course cut down the softer strata more rapidly than

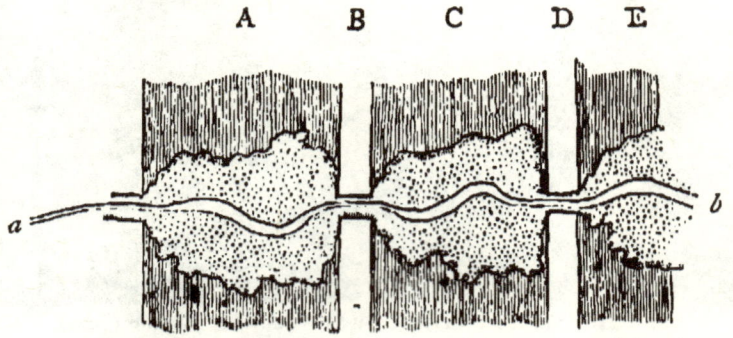

Fig. 46.—Diagram showing the course of a river through hard and soft strata.

the harder ones; each ridge of harder rock will therefore form a dam and give rise to a rapid or cataract. In cases such as these each section of the river has for a time a "regimen" of its own.

Suppose for instance a river $a\,b$ (Fig. 46) running

across the strike of several layers differing in hardness, *A, C, E,* being soft while *B, D* are tough or hard. In such a case the valley will widen out at *A, C, E.* Speaking generally we may say that the depth of the valley is mainly due to the river erosion, the width to weathering. Thus the Urseren Thal on the St. Gotthard, the broad stretches of valley at Liddes, and at Chable on the Dranse (Valais), are due to the more readily disintegrated Carboniferous or Jurassic strata. On the other hand, the depth of the valley will tend to arrive at the regular "regimen" (Fig. 47), and must in any case follow its normal course; but the width will depend on the destructibility of the strata. Even however the hardest rocks will give way in time, so that the inclination of the sides will depend on the hardness of the rocks and the age of the valley. Other things being equal, the older the valley the gentler will be the slopes of the sides.

Flat valley plains may be formed either by rivers or in a lake, and the surface view is the same in either case. The inner structure, however, as shown in a section, is very different. A river plain shows irregular, lenticular masses of gravel and sand. A stream running into a lake deposits fine mud in gently inclined layers, but as soon as it comes to the water's edge the coarser gravel rolls downwards forming a steeper slope.

The great Swiss valleys are of immense antiquity; the main ones were coeval with the mountains, and date back to the formation of the Alps themselves. Many indeed were even deeper in glacial times, having been to a great extent filled up by glacial deposits. Penck states that longitudinal are generally older than cross valleys. It seems to me, on the contrary, that they would as a rule have begun simultaneously. No doubt, however, there were many exceptions. The Dranse was probably an older river than the Upper Rhone. The Rhine below Basle runs in a comparatively recent depression. The greater number of the upper Swiss valleys must however date back to Miocene and some even to Eocene times, when rapid rivers were bringing down immense quantities of gravel from the slopes of the slowly rising Alps.

CHAPTER VII.

ACTION OF RIVERS.

ALTHOUGH the elevation of the Swiss Alps is the result of geological causes, the present configuration of the surface is mainly due to erosion and denudation. It is indeed impossible to understand the physical geography of any country without some knowledge of the action of water, and especially of rivers.

The velocity of a stream depends partly on the inclination of its bed, and partly on the volume of

FIG. 47.- Final Slope of a River.

water; if then we study an ancient river which has passed the stormy period of childhood, and forced its way through the obstacles of middle life, so that its waters run with approximately equal rapidity, we shall find that the slope diminishes from its source to the sea or lake into which it falls, with some such curve as in Fig. 47.

Such a river is said to have attained its "regimen," and this is the goal to which all rivers are striving to arrive.

The course of a river may be divided into three stages, which may be, and often are, repeated several times, viz.:—

1. Deepening and widening (the torrent).
2. Widening and levelling (the river proper).
3. Filling up (the delta).

and every part of a river in the second stage has passed through the first, every one in the third through the other two.

In the Valais the Upper glacier is a valley in the second stage, the ice-fall in the first; the plain from the foot of the fall to the Hotel in the second, from the Hotel to near Oberwald in the first; from Oberwald nearly to Niederwald in the second, from Niederwald to rather beyond Viesch in the first; then on to Brieg in the second, and from St. Maurice to Villeneuve in the third.

First Stage.

In the first phase the river has a surplus of force. It may be called a torrent. It cuts deeper and deeper into its valley, and carries away the mud and stones to a lower level. The sides are steep, as steep indeed as the nature of the material will permit, and the valley is in the shape of a V with little, if any, flat bottom. The water moreover continually eats back into the higher ground. The character of the valley depends greatly on that of the strata, being narrower where they are hard and tough, broader on the contrary where they are soft, so that they crumble more easily into the stream under the action of the weather. Fig. 46.

In several cases indeed the Swiss rivers run through gorges of great depth, and yet very narrow, even in some places with overhanging walls. The Via Mala, which leads from the green meadows of Schams (Sexamniensis, from its six brooks) to Thusis, is about five miles in length with a depth of nearly 500 metres, and very narrow, in one place not more than 9 to 15 metres in breadth.

The gorges of the Aar, of the Gorner, of the Tamina at Pfäffers, of the Trient, have a similar character. These were formerly supposed to be fissures due to upheaval. They none of them however present a trace of fracture, marks of water action can in places be seen from the base to the summit, and there can be no doubt that they have been cut through by the rivers.

In certain cases indeed we have conclusive

evidence. Some of these gorges are left at times quite dry, and it is easy then to see that the rock is continuous from side to side. The tunnels on the St. Gotthard line pass no less than six times under the Reuss, and there is no trace of a fault.

It may, I think, be said that the theory which attributed these gorges to a split in the rock is now definitely abandoned.

Of course, however, there are some cases in which the courses of streams have been determined by lines of fault and fracture.

Second Stage.

The second stage commences where the inclination becomes so slight that the river can scarcely carry away the loose material brought from above, or showered down from the sides, but spreads it over the valley, in which it wanders from side to side, and which it tends continually to widen. Hence unless they are confined by artificial embankments, such rivers are continually changing their course, keeping however within the limits of the same valley. The width of the valley moreover depends on its age, as well as on the size of the river and the character of the rock.

If we imagine a river running down a regularly

inclined plane in a more or less straight line, any inequality or obstruction, or the entrance of a side stream, would drive the water to one side, and when once diverted it would continue in the new direction, until the force of gravity drawing the water in a straight line downwards equalled that of the force tending to divert its course. Hence the radius of the curves will follow a regular curve-law depending on the volume of water and the angle of inclination

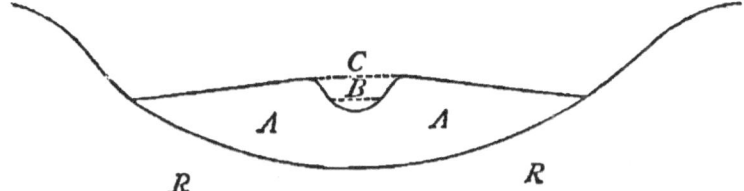

FIG. 48.—Diagrammatic Section of a Valley (exaggerated). *R R*, Rocky basis of valley; *A A*, Sedimentary strata; *B*, Ordinary level of river; *C*, Flood level.

of the bed. If the fall is ten feet per mile and the soil homogenous, the curves would be so much extended that the course would appear almost straight. With a fall of 1 foot per mile the length of the curve is, according to Fergusson, about six times the width of the river, so that a river 1000 feet wide would oscillate once in 6000 feet. This is an important consideration and much labour has been lost in trying to prevent rivers from following their natural laws of oscillation. But rivers are very true to their own laws, and a change at any part is continued both upwards and downwards, so that a new oscillation in any place cuts its way through the whole plain of the river both above and below.

If the river has no longer a sufficient fall to enable it to carry off the materials it brings down, it gradually raises its bed (Fig. 48), hence in the lower part of their course many of the most celebrated rivers—the Po, the Nile, the Mississippi, the Thames, etc.—run upon embankments, partly of their own creation.

The Reno, the most dangerous of all the Apennine rivers, is in some places more than 30 feet above the adjoining country. Rivers under such conditions, when not interfered with by Man, sooner or later break through their banks, and, leaving their former bed, take a new course along the lowest part of their valley, which again they gradually raise above the rest.

Along the valley of the Rhone from Visp down to the Lake of Geneva there is often a marsh on one side of the valley, sometimes on both, the existence of which may be thus explained.

This is the second stage.

Third Stage.

Finally, when the inclination becomes too small the stream cannot carry farther the stones and mud which it has brought down, and spreads them out in the form of a fan, forming a more or less flat cone or delta—a cone if in air, a delta if under water; and the greater the volume of water, the gentler will the slope be, so that in great rivers it becomes almost imperceptible. At this part of its course, the stream instead of meandering, will tend to divide into several branches.

Cones and deltas are often spoken of as if they were identical. The surface and slope are indeed similar, but the structure of a delta formed under water (see p. 186) is by no means the same as that of a cone formed in the air.

Deltas have generally a very slight inclination, so far as the surface is concerned, while the layers below stand at a greater inclination. Most of the Swiss Lakes are being gradually filled up by the deposits of rivers. The Lake of Geneva once extended far up the Rhone Valley to St. Maurice if not to Brieg. It presents also a very typical delta at the mouth of the Dranse near Thonon. Between Vevey and Villeneuve are several such promontories, each marking the place where a stream falls into the lake.

Where lateral torrents fall into a main valley the rapidity of the current being checked, their power of transport is diminished, and similar "river cones" are formed. A side stream with its terminal cone, when seen from the opposite side of the valley, presents the appearance shown in Fig. 49, or, if we are looking down the valley, as in Fig. 50, the river being often driven to one side of the main valley, as, for instance, is the case in the Valais near Sion, where the Rhone is (Fig. 51 p. 199) driven out of its course by, and forms a curve round, the cone formed by the River Borgne.

The river cones are, in many cases, marked out by the character of the vegetation. "The Pines enjoy the stony ground particularly, and hold large meetings upon it, but the Alders are shy of it, and, when it has come to an end, form a triumphal procession all round its edge, following the convex line."*

The magnitude of these "river cones" depends on the amount and character of the materials brought into the main valley, and on the power of the river to carry them off. The felling of

* Ruskin, *Modern Painters*, vol. IV.

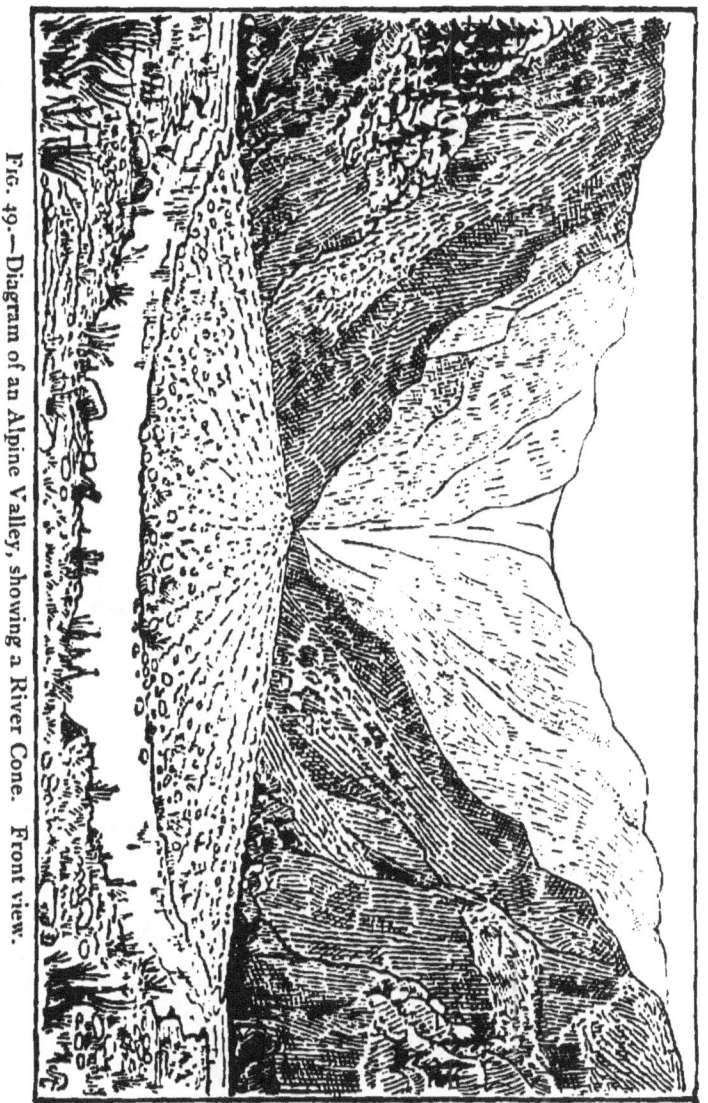

Fig. 49.—Diagram of an Alpine Valley, showing a River Cone. Front view.

ACTION OF RIVERS. 197

Fig. 50.—Diagram of an Alpine Valley, showing a River Cone. Lateral view.

forests, for instance, in a lateral valley will considerably increase the erosive power of the stream, and the amount of material brought down. Rocks which yield readily to the action of weather and water will naturally supply most material, and give rise to the largest cones, especially if they form hard pebbles. On the other hand, the Flysch, which, as a rule, exercises little resistance, does not produce such important cones as might be expected, because it disintegrates into fine particles which are easily washed away. The Cargneule, on the contrary, produces large cones, because it breaks up readily, but into hard pieces.

Such cones sometimes raise the bed of the valley and dam back the water, and thus form a marshy and unhealthy tract. Thus in the Upper Valais below Oberwald is a succession of such cones, one succeeding another, and with more or less marshy ground between them. At Münster there is a fine cone, and further down are many others at intervals. The two largest are those of the Illgraben at Leuk, and the Chamoson at the mouth of the Losentze, both of which raise the level of the valley above several feet. That of the Borgne (Fig. 51), near Sion, drives the river to the foot of the opposite mountain.

When at length a river has so adjusted its slope that it neither deepens its bed in the upper portion

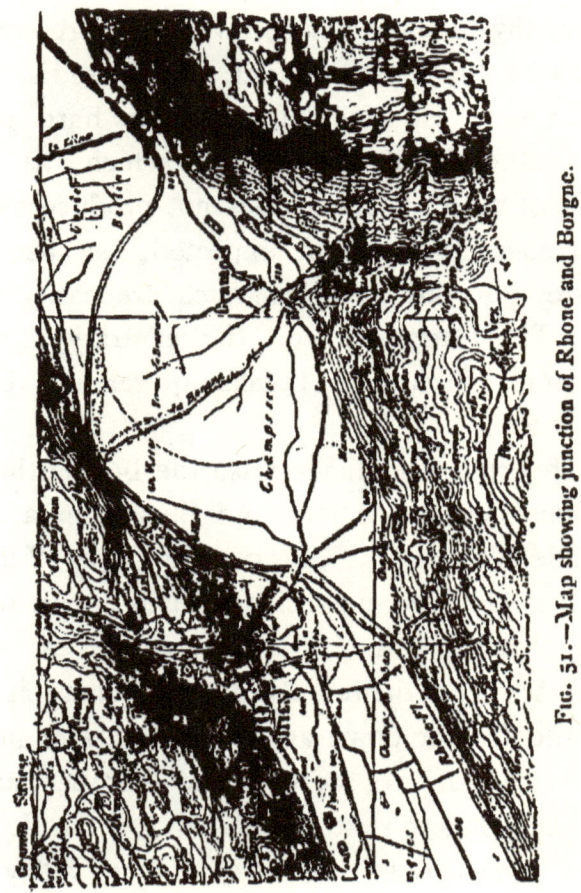

Fig. 51.—Map showing junction of Rhone and Borgne.

of its course, nor deposits materials, it is said to have acquired its "regimen" (Fig. 47 p. 188), and in such a case the velocity will be uniform. The enlargement of the bed of a river is not, however, in proportion to the increase of its waters as it approaches the sea. Other things being equal, a river which increases in volume, increases in velocity; the "regimen" therefore would be destroyed, and the river would again commence to eat out its bed.

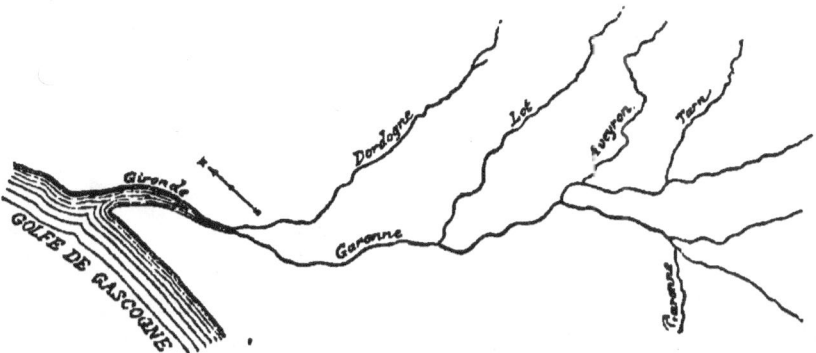

Fig. 52.—River system of the Garonne.

Hence, if rivers enlarge, as for instance owing to any increase in territory, the slope diminishes.

The above figure (Fig. 52) gives a sketch map, and Fig. 53 represents the profiles, of the principal rivers in the valley of the Garonne, and it will be seen that the larger the river the gentler is the slope.

At present many of the smaller Swiss streams are eating into their cones and endeavouring to

ACTION OF RIVERS. 201

flatten them, owing perhaps to the gradual enlargement of the gathering-grounds.

These cones are favourite sites for villages, which are thus raised and placed above the range of ordinary floods. The loose materials of the upper part of the cone, moreover, absorb water freely in the upper part, which is filtered, and emerges in clear springs lower down. Thus arise many of the fountains in such villages.

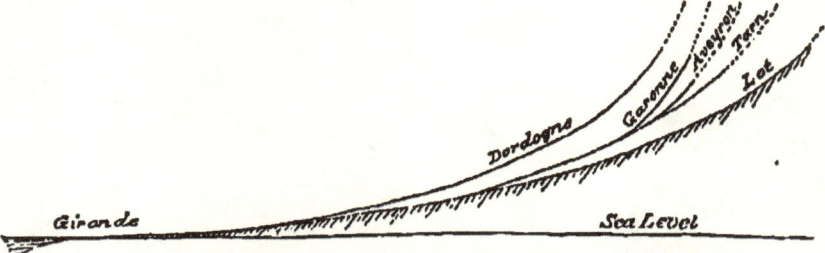

FIG. 53.—Slopes of the Garonne and its principal affluents.

Now let us suppose that the force of a river is again increased, either by a fresh elevation, or locally by the removal of a barrier, or by an increase in volume owing to an addition of territory, or greater rainfall, it will then again cut into its own bed, deepening the valley, and giving rise to a rapid, which will creep gradually up the valley, receding of course more rapidly where the strata are soft, and lingering longer at any hard ridge.

The old plain of the valley will form a more or

less continuous terrace above the new course. Such old river terraces may be seen in most valleys; often indeed several, one above another. The upper terraces being generally cut in the rock, the lower ones in river deposits or fallen debris.

It has been sometimes supposed that these ter-

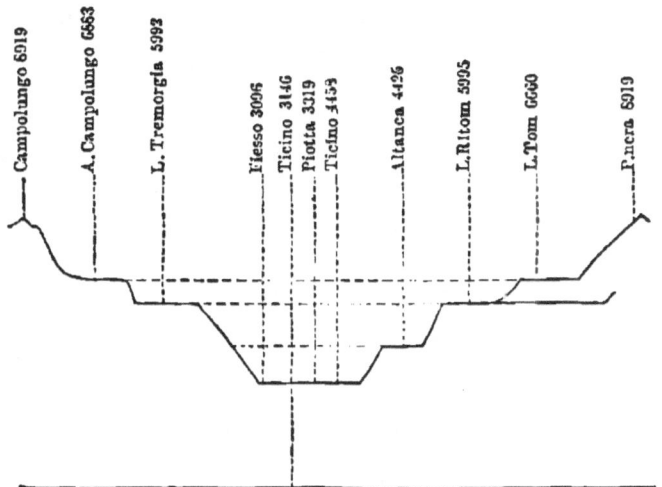

FIG. 54.—Section across the valley of the Ticino. On the left from the Ticino to Campolungo; on the right by Altanca to P.nera.

races indicate a greater volume of water in ancient times, sufficient indeed to fill up the whole valley to that depth. It must be remembered, however, that the terrace was formed before the lower part of the valley was excavated.

Fig. 54 is a section across the valley of the Ticino, a short distance below Airolo. It shows two high terraces on which the Lakes Tom and Ritom are respectively situated, and which correspond to those of Campolungo and Tremorgia on the other (W.) side of the valley. Below them is another terrace at a height of 1350 metres, on which Altanca stands. This terrace can be traced for some distance, and bears a series of villages—Altanca, Ronco, Beggio, Catto, Osco, etc. In the valley of the Ticino there is a

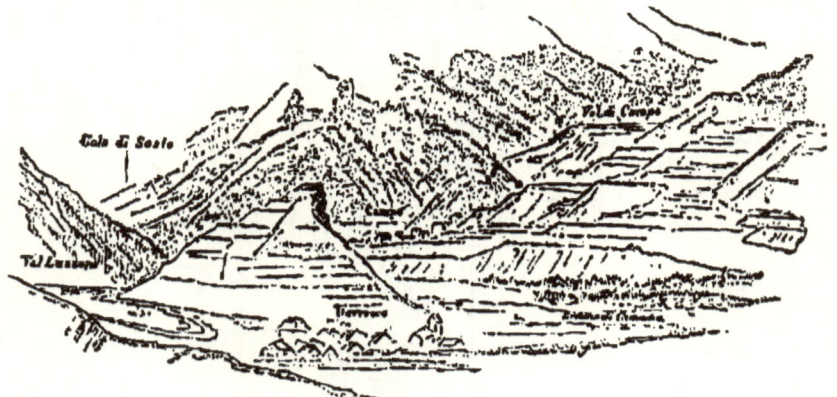

FIG. 55.—River Terraces in Val Camadra.

second series of still more important towns, at, or at least little above, the present river bed, but in other cases, as, for instance, along the Plessur, which falls into the Rhine at Chur, the present river bed is quite narrow, and the villages are on an old river terrace high above the present water level.

Fig. 55 represents a group of river terraces in the Val Camadra.

In each river system the terraces occupy corresponding levels, but in different systems they have no relation to one another. They afford, as we shall see in the next chapter, valuable evidence as regards the former history of rivers.

Hitherto I have assumed that the river deepens its bed vertically. This is not, however, always the case. If the strata are inclined the action of the

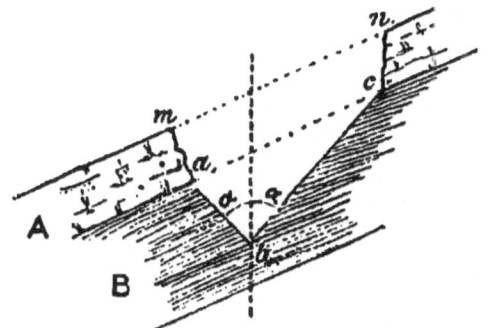

FIG. 56.—Diagram of River Valley.

water will tend to follow the softer stratum, as for instance, in the following diagram, where A represents a harder calcareous rock overlying a softer bed B.

The enormous amount of erosion and denudation which has taken place may be estimated from the fact that terraces can still be traced in some cases at a height of 3000 metres above the present river beds.

As we approach their source, valleys become

steeper and steeper. In some cases, and especially in calcareous districts, the valleys end in a precipitous, more or less semicircular "Cirque." Springs rising at the foot of such escarpments are known as Vauclusian, from the celebrated and typical instance at Vaucluse.

Another interesting point brought out by the study of Swiss rivers, is that just as in Geology, though there have no doubt been tremendous cataclysms, still the main changes have been due to the continuous action of existing causes; so also in the case of rivers, however important the effects produced due to floods, still the configuration of river valleys is greatly due to the steady and regular flow of the water.

Floods may be divided into two classes, (1) those due to the bursting of some upper reservoir, such, for instance, as the great flood of the Dranse de Bagnes in 1818, due to the outburst of the lake, which had been dammed back by the glacier of Giétroz, or the more recent flood of St. Gervais owing to the bursting of a subglacial reservoir in the little Glacier de Tête Rousse which rushed down the valley in the dead of the night, in a few minutes swept away the Baths, and drowned most of the visitors; and (2) those due to heavy rains. No one can travel much in Switzerland without seeing the great precautions taken to confine the rivers within certain limits. In fact, what we call the river bed, is rather the low-water channel, and the whole bottom of the valley would, but for these precautions, be covered during any considerable flood. Egypt itself is the river bed of the Nile during the autumn flood.

GIANTS' CALDRONS.

These are more or less circular cavities, often somewhat raised in the centre. They sometimes attain a considerable size—as much as 8 metres in diameter and 5 in depth. There is a very fine group at Lucerne, where they are known as the "Jardin du Glacier." They have been excavated in the rock by blocks of harder stone being whirled round by the action of water. Some of them no doubt, and certainly those at Lucerne, were formed under glaciers, at the foot perhaps of a "moulin," but I believe that as a rule they were formed in streams.* Several have recently been discovered at the Maloja; there are some fine specimens also near Servoz in the valley of the Arve. Renevier points out that such caldrons can be seen actually in process of formation in some of the existing rivers, as, for instance, near

* Favre, *Rech. Geol.*, vol. I.

the junction of the Rhone and the Valorsine below Geneva. These, however, will be destroyed as erosion continues. Surprise is sometimes expressed that Giants' Caldrons occur where no stream now flows. But it is just to this fact that they owe their existence. If the river had not changed its course they would long since have been destroyed.

Before closing this chapter I must say a few words about subterranean streams. These occur mainly in porous rocks, such as those of the Jura. The most considerable of these partly subterranean rivers is the Orbe, which rises originally in a little French lake, Les Rousses, traverses two others on Swiss territory, the Lake de Joux, and that of Brenet, and then disappears suddenly in the ground at the foot of a high cliff, reappearing again at a distance of 3 km. near Vallorbes.

Summing up this chapter we may say that as soon as any tract of land rose out of the sea, the rain which fell on the surface would trickle downwards in a thousand rills, forming pools here and there, and gradually collecting into larger and larger streams. Whenever the slope was sufficient, the water would begin cutting into the soil and carrying it off to the sea. This action would, of course, differ in rapidity according to the slope and hardness of the ground. The character of the valley would depend greatly on the nature of the strata, being narrow where they were hard and tough; broader, on the contrary, where they were soft, so that they crumbled readily into the stream, or where they were easily split by the weather. Gradually the stream would eat into its bed until it reached a certain slope, the steepness of which would depend on the volume of water. The erosive action would then cease, but the weathering of the sides and consequent widening would continue, and the river would wander from one part of the valley to another, spreading the materials and forming a river plain. At length, as the rapidity still further diminished, it would no longer have sufficient power even to carry off the materials brought down. It would form therefore a cone or delta, and instead of wandering would tend to divide into different branches.

When we look at some great valley of denudation and the comparatively small river which flows through it, we may deem it almost impossible that so great an effect can be due to so small a cause. We find, however, every gradation from the little gully cut out by the last summer shower up to the great Cañon of Colorado. We have to consider not only the flow of the water, but the lapse of time, and remember that our river valleys are the work of ages. Moreover, even without postulating any greater rainfall in former times, we must bear in mind that we are now looking at rivers which have attained or are approaching their equilibrium; they are comparatively steady, and even aged; so that we cannot measure their present effect by that which they produced when they possessed the energy and impetuosity of youth.

From this point of view the upper part of a river valley is peculiarly interesting. It is a beautiful and instructive miniature. The water forms a sort of small-meshed net of tiny runnels. We can as it were surprise the river at its very commencement; we can find streamlets and valleys in every stage, a quartz pebble may divert a tiny stream, as a mountain does a great river; we find springs and torrents, river terraces and waterfalls, lakes and deltas in the space of a few square metres, and changes pass under our eyes which on a larger scale require thousands of years.

And as we watch some tiny rivulet, swelling gradually into a little brook, joined by others from time to time, growing to a larger and larger torrent, then to a stream, and finally to a great river, it is impossible to resist the conclusion gradually forced

Scenery of Switzerland. I.

upon us, that, incredible as it must at first sight appear, even the greatest river valleys, though their origin may be due to the original form of the surface, owe their present configuration mainly to the action of rain and rivers.

Note.—Throughout western Europe a large proportion of the river names fall into three groups.

From the Old German Aha, Celtic Uisge, Gaelic Oich, Latin Aqua (Water), softened into the French Eau, we have the Aa, Awe, Au, Avon, Aue, Ouse, Oise, Grand Eau, Aubonne, Oich, Ock, Aach, Esk, Uisk, etc.

From the Celtic Dwr (Greek ὕδωρ), we have Oder, Adour, Thur, Dora, Douro, Doire, Durance, Dranse, Doveria, etc.

From the Celtic Rhin, or Rhedu, to run (Greek ῥέω), we have the Rhine, Rhone, Reuss, Reno, Rye, Ray, Raz, etc.

The names Aa and Drance or Dranse are so common in Switzerland that it is necessary to specify them by some further description, such as the Engelberger Aa, the Aa of Alpnach, the Milch Aa, Hallwyler Aa, Wäggithaler Aa, etc.

The Drance which falls into the Lake of Geneva near Thonon is perhaps the Drance *par excellence,* but in the same river system we have also the Drance de Bagne, the Drance d'Entremont, and the Drance de Ferret.

In the case of the Rhine also there is the Vorder Rhein, Mittel Rhein, Hinter Rhein, Oberhalbstein Rhein, Averser Rhein, Safien-Rhein, etc.

CHAPTER VIII.

DIRECTIONS OF RIVERS.

THE general direction of the river-courses in any country is determined in the first instance by the configuration of the surface at the time of its becoming dry land. The least inequality in the surface would affect the first direction of the streams, and thus give rise to channels, which would be gradually deepened and enlarged. They are, however, in many cases materially modified by subsequent changes of relative level, and by the results of erosion, which acts of course much more rapidly on some strata than on others. It is as difficult, however, for a river as it is for a man to get out of a groove.

If we imagine a district raised in the form of a regular dome, the rivers would radiate from the summit in all directions. The lake district in the north of England; the Plateau of Lanneme-zan in the south of France, and the Ellsworth Arch in the Henry Mountains,* offer us approximations to such

* See Gilbert, *Geology of the Henry Mountains.*

a condition. It seldom happens, however, that the case is so simple, and the lines of rivers offer many interesting problems, which are by no means easy to solve.

As already mentioned (*ante*, p. 174), the Swiss rivers follow two main directions, at right angles to one another, namely, S.W. by N.E. and N.W. by S.E. The first follows the strike of the strata. The explanation of the second is not so simple. The probable cause, however, which has determined the two main directions of the Swiss rivers has been already suggested (*ante*, p. 181).

The principal Swiss rivers must be of great antiquity. Some of the streams in the eastern and central parts of the Alps probably commenced even in Eocene times. The Nagelflue was brought down from the mountains by rivers which probably occupied the upper parts of the valleys of the Aar, Reuss, etc.

Nevertheless there have been great changes in the courses of the Swiss rivers. These are ascribable to four main causes:—First, it must be remembered that streams are continually eating back into the hills. In many cases they cut completely through them, and if the valley into which they thus force their way is at a higher level, they carry off the upper waters; Secondly, later earth movements in many cases changed the course of the rivers; Thirdly, they have in many cases been diverted by masses of glacial deposits; and Fourthly, the summit ridge of the Alps is slowly retreating northwards, which affects the river system of all the upper districts.

In the great Swiss plain the country slopes on the whole northwards from the Alps, so that the lowest part is that along the foot of the Jura. Hence (Fig. 42) the main drainage runs along the line from Yverdun to Neuchâtel, down the Zihl to Soleure, and then along the Aar to Waldshut. The Upper Aar, the Emmen, the Wigger, the Suhr, the Wynen, the lower Reuss, the Sihl, and the Limmat, besides several smaller streams, running approximately parallel to one another—N.N.W., and at a right angle with the main axis of elevation, all join the Aar from the south, while on the north it does not receive a single tributary of any importance.

On the south side of the Alps again, and for a corresponding reason, all the great affluents of the Po—the Dora Baltea, the Sesia, the Ticino, the Olonna, the Adda, the Adige, etc., come from the north, and run S.S.E. from the axis of elevation to the Po.

Indeed, the general slope being from the ridge of the Alps towards the north, most of the large affluents of rivers running in longitudinal valleys fall in on the south, as, for instance, those of the Isère from Albertville to Grenoble, of the Rhone from its source to Martigny, of the Vorder Rhine from its source to Chur, of the Inn from Landeck to Kufstein, of the Enns from its source to near Admont, of the Danube from its source to Vienna, and, as just mentioned, of the valley from Yverdun to Waldshut. Hence also, whenever the Swiss rivers running east and west break into a transverse valley, as the larger ones all do, and some more than once, they invariably, whether originally running east or west, turn towards the north.

But why has the plain of Switzerland this slope? Why is it lowest along the wall of the Jura? As has been already pointed out, this part of Switzerland was formerly a sea, which was gradually filled by river deposits. It is indeed a great "cone" due to many rivers which flowed down from the rising Alps. This being so, the general slope is naturally up to, and the lowest part is that farthest away from, the mountains.

In considering the courses of rivers it must be remembered that the strata situated below by no means always correspond with those at a higher level. Again it will sometimes happen that rivers follow a course which is very difficult to explain, because, in fact, it has no reference to the present configuration of the surface, but has been determined by the existence of strata which have now disappeared.

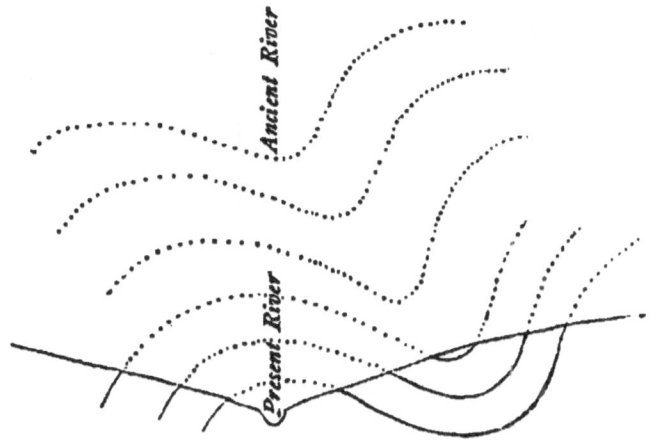

Fig. 57.—Diagram to illustrate a river now running in an anticlinal.

It often happens, for instance, that the rivers now run apparently on an anticlinal, and have a synclinal on one side (Fig. 57), as, for instance, the Rhine at Dissentis (see Fig. 134, vol. II. p. 191).

The folds, however, being inclined, it will be seen from the dotted lines (Fig. 57) that when the river began its labour it perhaps did run in the synclinal, but having cut its way directly downwards is now some way from it, and will diverge further and further as erosion proceeds.

It is a remarkable fact that great folds by no means always determine the watershed, but, on the contrary, rivers often cut through ranges of mountains.

Thus the Elbe cuts right across the Erzgebirge, the Rhine through the mountains between Bingen and Coblenz, the Potomac, the Susquehannah, and the Delaware through the Alleghanies. Even the chain of the Himalayas, though the loftiest in the world, is not a watershed, but is cut through by rivers in more than one place. The case of the Dranse will be alluded to further on. In these instances the rivers probably preceded the mountains. Indeed, as soon as the land rose above the waters, rivers would begin their work, and having done so, if a subsequent fold commenced, unless the rate of elevation exceeded the power of erosion of the river, the two would proceed simultaneously, so that in many cases the river would not alter its course, but would cut deeper and deeper as the mountain range gradually rose.

In some other cases where we speak of a river suddenly changing its direction, it would be more correct to say that it falls into the valley of another stream. Thus the Aar, below Berne, instead of continuing in the same direction, by what seems to have been its ancient course, along the broad valley now only occupied by the little Urtenenbach, suddenly turns at a right angle, falling into the valley of the Sarine, near Oltigen.

Take again the Rhone (Fig. 58). It is said to turn at a right angle at Martigny, but in reality it falls into and adopts the transverse valley, which properly belongs to the Dranse; for the Dranse is probably an older river and ran in the present course even before the origin of the Valais. This would seem to indicate that the Oberland range is not so old as the Pennine, and that its elevation was so gradual that the Dranse was able to wear away a passage as the ridge gradually rose. After leaving the Lake of Geneva the Rhone follows a course curving gradually to the south, until it falls into and adopts a valley which properly belongs to the Valserine, and afterwards another belonging to the little river Guiers; it subsequently joins the Ain, and finally falls into the Saône. If these valleys were attributed to their older occupiers, we should therefore confine the name of the Rhone to the portion of its course from its source to Martigny.

From Martigny it invades successively the valleys of the Dranse, Valserine, Guiers, Ain, and Saône. In fact, the Saône receives the Ain, the Ain the Guiers, the Guiers the Valserine, the Valserine the Dranse, and the Dranse the Rhone. This is not a mere question of names, but also one of antiquity. The Saône, for instance, flowed past Lyons to the

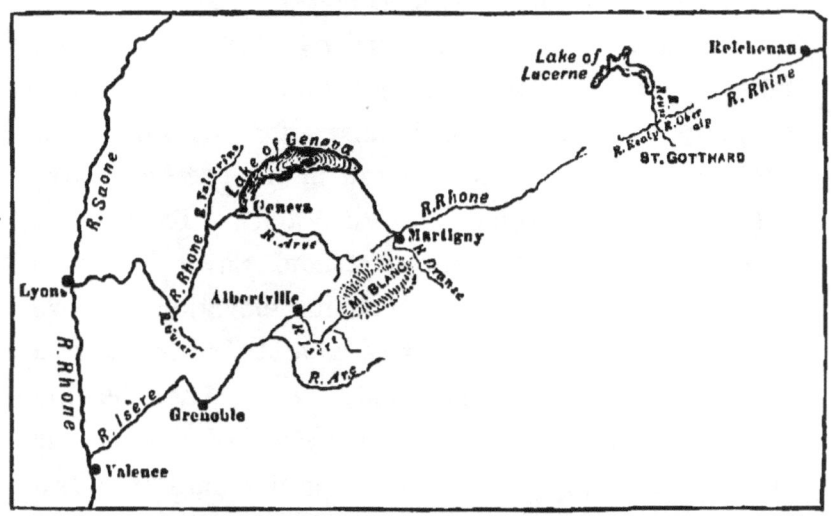

Fig. 58.—Sketch map of the Rhone and its tributaries.

Mediterranean for ages before it was joined by the Rhone. In our nomenclature, however, the Rhone has swallowed up the others. This is the more curious from the fact that of the three great rivers which unite to form the lower Rhone, namely, the Saône, the Doubs, and the Rhone itself, the Saône brings for a large part of the year the greatest volume of water, and the Doubs has the longest course.

We will now consider some of the cases in which

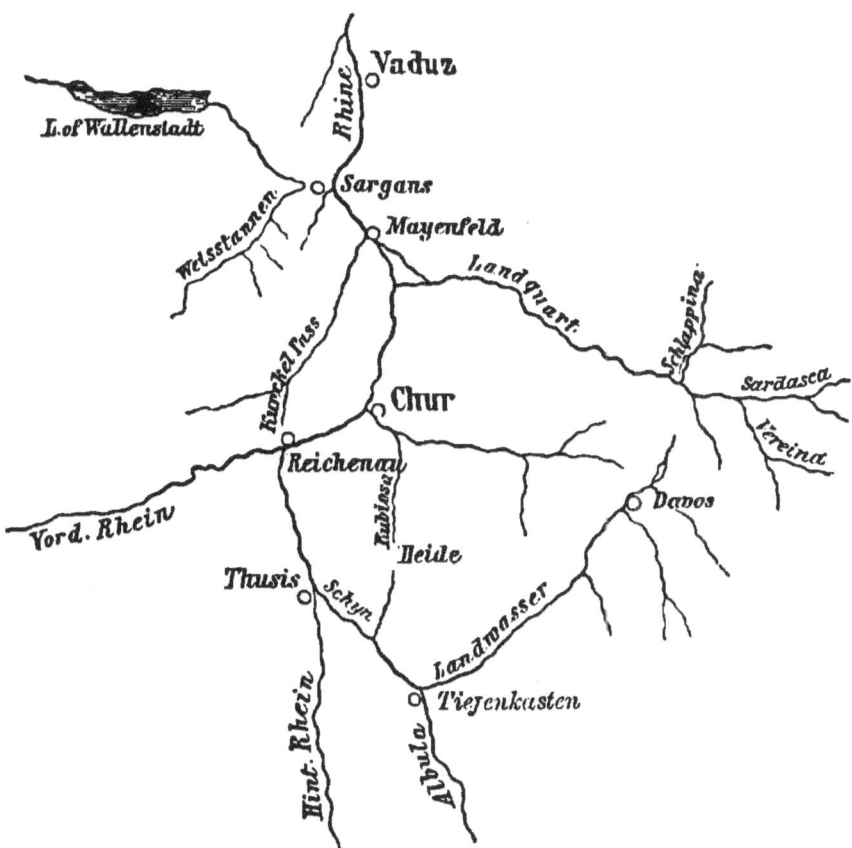

FIG. 59.—River system round Chur, as it is.

Swiss rivers have altered their courses. In some of these the change of direction is doubtless due to the fact that some stream at the lower level, or with a greater fall, has eaten its way back, and so tapped the higher valley.

Rivers, indeed, have their adventures and vicis-

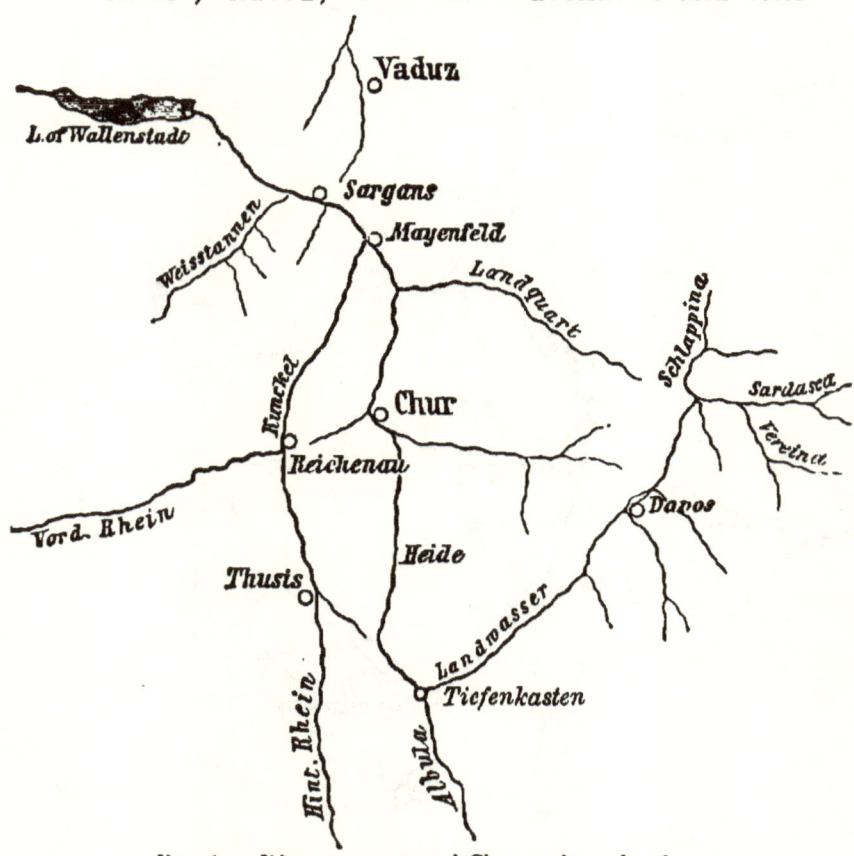

FIG. 60.—River system round Chur, as it used to be.

situdes, their wars and invasions. Take, for instance, the Upper Rhine (Fig. 59), of which we have a very interesting account by Heim. It is formed of three main branches, the Vorder Rhine, the Hinter Rhine, and the Albula. The two latter, after meeting near Thusis, unite with the Vorder Rhine at Reichenau, and run by Chur, Mayenfeld, and Sargans into the Lake of Constance at Rheineck. At some former period, however, the drainage of this district was very different.

The Vorder and Hinter Rhine united then, as they do now, at Reichenau, but at a much higher level, and ran to Mayenfeld (Fig. 60), not by Chur, but by the Kunkels Pass to Sargans, and so onwards not to the Lake of Constance, but to that of Zürich. The Landwasser at that time rose in the Schlappina Joch, and after receiving as tributaries the Vereina and the Sardasca, joined the Albula, as it does now at Tiefenkasten; but instead of going round to meet the Hinter Rhine near Thusis, the two together travelled parallel with, but at some distance from, the Hinter Rhine, by Heide to Chur, and so to Mayenfeld.

As we look up from Tiefenkasten towards Heide and the Parpan Pass it seems almost incredible that the Oberhalbstein Rhine can ever have taken that course. I give therefore (Fig. 61) the following profile showing the old river terrace, but with the height exaggerated in comparison with the distance. This,

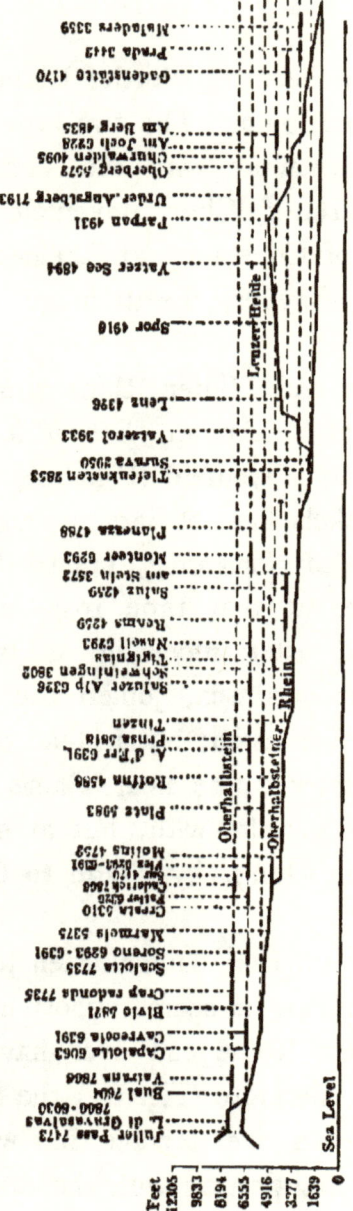

FIG. 61.—Longitudinal Profile of the Oberhalbstein Rhine and the Parpan Pass. 1=100,000.

however, does not affect the relative elevations. The dotted lines follow the natural slope of a river, and the strengthened parts show where portions of terrace still remain. It is obvious that before the ancient Schyn had cut its way up to Tiefenkasten the Oberhalbstein Rhine and the Landwasser flowed over the Parpan Pass, and not only flowed over it, but have cut it down some 610 metres, that is to say, when the river flowed over it with its natural regimen in relation to the valley it was at a height of 2200 metres, and has left a fragment of terrace at that height at Urder Angstberg, the Parpan itself being only 1500 metres.

In fact, the Parpan and Kunkels passes are deserted river valleys, showing on each side river terraces, and were obviously once the beds of great rivers, very different from the comparatively small streams which now run in their lower parts.

In the meanwhile, however, the Landquart stealthily crept up the valley, attacked the ridge which then united the Casanna and the Mädrishorn, and gradually forcing the passage between Dörfli and Klosters, invaded the valleys of the Schlappina, Vereina, and Sardasca, absorbed them as tributaries, detached them from their allegiance to the Landwasser, and annexed the whole of the upper province, which had formerly belonged to that river.

The Schyn also gradually worked its way upwards from Thusis till it succeeded in sapping the Albula, and carried it down the valley to join the Vorder Rhine near Thusis. In what is now the main valley of the Rhine above Chur, another stream ate its way back, and eventually tapped the main river at Reichenau, thus diverting it from the Kunkels and carrying it round by Chur.

It is possible that in the distant future the Landwasser may be still further robbed of its territory. The water of the Davos Lake, the Flüela, the Dischma, and the Kuhalpthal now take a very circuitous route to Chur, and it is not impossible that they may be captured and carried off by the Plessur.

At Sargans a somewhat similar process was repeated, with the addition that the material brought down by the Weisstannen, or perhaps a rockfall, deflected the Rhine, just as we have seen (*ante,* p. 198) that the Rhone was pushed on one side by the Borgne. The Rhone, however, had no choice, it was obliged to force, and has forced, its way over the cone deposited by the Borgne. The Rhine, on the contrary, had the option of running down by Vaduz to Rheineck, and has adopted this course.

The association of the three great European rivers —the Rhine, the Rhone, and the Danube—with the past history of our race, invests them with a singular fascination, and their own story is one of much interest. They all three derive part of their upper waters from the group of mountains between the Galenstock and the Bernardine, within a space of a few miles; on the east the waters now run into the Black Sea, on the north to the German Ocean, and on the west to the Mediterranean. But it has not always been so. Their head-waters have been at one time interwoven together.

The present drainage of Western Switzerland is very remarkable. If you stand on a height overlooking the valley of the Arve near Geneva, you see a semicircle of mountains—the Jura, the Vuache, the Voirons, etc., which enclose the west end of the Lake of Geneva; the Arve runs towards the lake, which itself opens out towards Lausanne, where a tract of low land alone separates it from the Lake of Neuchâtel and the valley of the Aar. This seems the natural outlet for the waters of the Rhone and the Arve. As a matter of fact, however, they escape from the Lake of Geneva at the western end, through the remarkable defile of Fort de l'Ecluse and Maupertuis, which has a depth of nearly 300 metres, and is at one place not more than 14 feet across. There are reasons, moreover, as we shall see presently, for considering the defile to be of comparatively recent origin. Moreover, at various points round the Lake of Geneva, remains of lake terraces show that the waters once stood at a level much higher than at present. One of these is rather more than 76 metres above the lake.

The low tract between Lausanne and Yverdun has a height of 76 metres (250 feet) only, and corresponds with the above-mentioned lake terrace. The River Venoge, which rises between Rolle and the Mont Tendre, runs at first towards the Lake of Neuchâtel, but near La Sarraz it divides; the valley continues in the same direction, and some of the water joins the Nozon, which runs to the Lake of Neuchâtel at Yverdun; but the river itself turns sharply to the south, and falls into the Lake of Geneva to the east of Morges.

It is probable, therefore, that when the Lake of Geneva stood at the level of the 76 metres terrace the waters ran out, not as now at Geneva and by Lyons to the Mediterranean, but near Lausanne by Cissonay and Entreroches to Yverdun, and through the Lake of Neuchâtel into the Aar and the Rhine.

But this is not the whole of the curious history.

At present the Aar makes a sharp turn to the west at Waldshut, where it falls into the Rhine, but there is some reason to believe that at a former period, the river continued its course eastward to the Lake of Constance, by the valley of the Klettgau, as is indicated by the presence of gravel beds containing pebbles which have been brought, not by the Rhine from the Grisons, but by the Aar from the Bernese Oberland, showing that the river which occupied the valley at that time was not the Rhine but the Aar. It would seem also that at one time the Lake of Constance stood at a considerably higher level, and that the outlet was, perhaps, from Friedrichshafen to Ulm, along what are now the valleys of the Schussen and the Ried, into the Danube.*

The River Aach, though a tributary of the Rhine, still derives its head-waters from the valley of the Danube. A part of the water of the Danube sinks into fissures in the Jurassic rocks at Immendingen, and makes its appearance again as copious springs at Aach, from whence they flow into the Lake of Constance near Rudolphzell.

Thus the head-waters of the Rhone appear to have originally run between Morges and Lausanne

---

* Du Pasquier, *Beitr. z. Geol. K. d. Schw.*, L. XXXI.

and to the Lakes of Neuchâtel and Constance into the Danube, and so to the Black Sea. Then, after the present valley was opened between Waldshut and Basle, they flowed by Basle and the present Rhine, and after joining the Thames, over the plain which now forms the German Sea into the Arctic Ocean

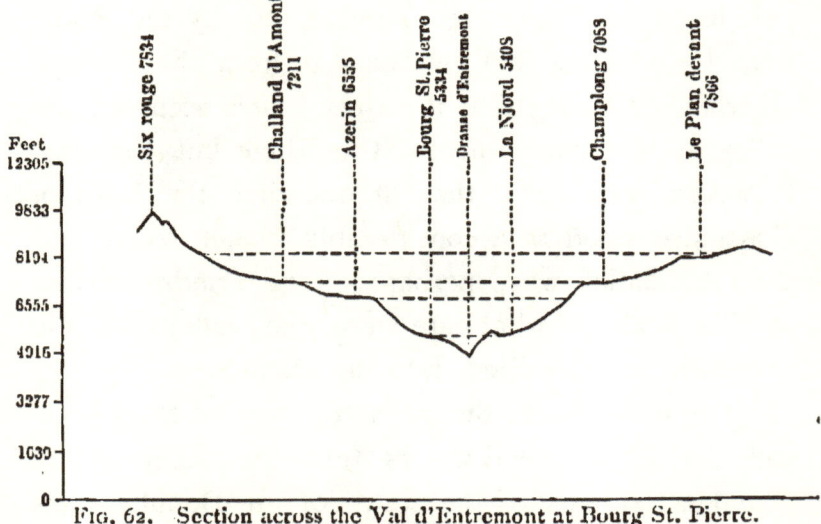

FIG. 62. Section across the Val d'Entremont at Bourg St. Pierre.
1=100,000.

between Scotland and Norway. Finally, after the opening of the passage at Fort de l'Ecluse, by Geneva, Lyons, and the valley of the Saône, to the Mediterranean.

In the upper parts of the district there have also been some changes.

Fig. 62 shows the river terraces on the Dranse

d'Entremont, near Bourg St. Pierre, where the Society for the protection of Alpine plants have established a very interesting Alpine garden, and (Fig. 63) those further down the valley near La Douay.

The uppermost of these terraces is at a height of 2200 metres. The col leading to the Vallée de

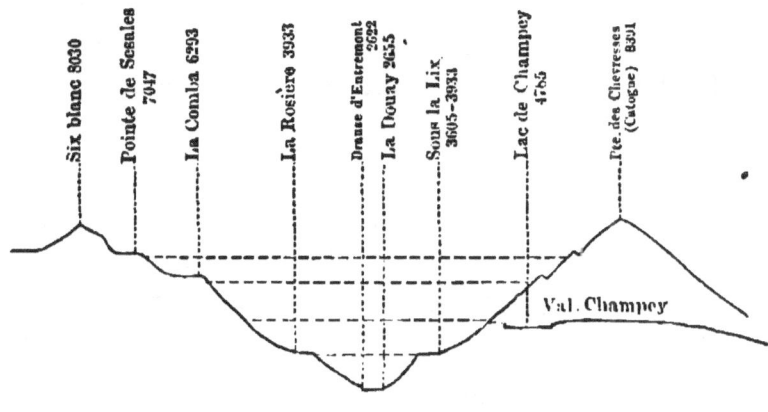

FIG. 63.— Cross section of the Valley of the Dranse, between the Valley of Champey, Sembranchier and Orsières.

Champey is at a height of about 1500 metres, and until the river had reached a lower level than this, the waters of the Dranse followed what the map shows was their natural course down the Vallée de Champey. Eventually, however, the Orsières branch of the Dranse de Bagne cut its valley back and carried off the upper waters to join the Dranse de

Bagne at Sembranchier. This was facilitated by the comparative softness of the Jurassic strata and Grey Schists, while the Vallée de Champey is in Protogine, Felsite, and Porphyry, which offered a much greater resistance to the action of the water.*

The Trient also has changed its course. Originally it ran over the Col de la Forclaz down to Martigny. In this case the change is due, not to any difference in the hardness of the rock, but to the greater fall, and consequently greater erosive power, of the Eau Noire.

It would also seem that some of the Vaud and Friburg rivers must be older than the final elevation of the mountains at the north-east end of the Lake of Geneva. Gillieron points out that the Broye, the Mionnaz, the Flon, and I may add the Sarine, from Sarnen to below Château d'Oex, run towards the Lake of Geneva, until they are stopped by the mountains between Chatel St. Denis and the Rocher de Naye, and forced to return northwards.

There is also one important change which applies to the whole crest of the Alps.

Watersheds are at first determined by the form of the earliest terrestrial surface, and if the slopes in

---

* Bodmer, p. 21.

each side are equal they will be permanent; on the other hand if, as in the Alps, one side is much steeper than the other, it will be worn back more rapidly. Hence the whole crest of the Alps is, though of course very slowly, moving northwards. This is specially marked in the case of the Engadine (see vol. II. p. 240).

These changes and struggles have by no means come to an end. In some cases we can already foresee future changes. For instance, the Nolla, which falls into the Hinter Rhine at Thusis, is rapidly eating back into the mountains near Glas, and in, geologically speaking, a comparatively short time it will probably invade the Valley of the Versam, carry off its upper feeders, and appropriate the waters from the upper valley. So rapidly is the change progressing that after even a few hours' rain the Nolla becomes quite black. In its upper part the Bündnerschiefer is saturated with water, and reduced almost to a black mud. The ground may be said to be continually in slow movement down to the valley, and the houses of Glas and Tschappina have to be continually repaired. Some have moved as much as 60 metres downwards in thirty years.

Age of Rivers.

It follows from these considerations not only that some Swiss rivers are of comparatively recent origin, while others date back to very great antiquity, but that different parts of what is now considered a single river are of very different ages and have a very different history.

The southern part of the Central Alps are supposed to have been first raised above the waters, and to have formed an Island in Eocene times, to which therefore some of the head-waters date back. It is, however, clear that the rivers crossing the Miocene deposits of Central Switzerland cannot have commenced until after the Miocene strata had been raised and become dry land. In fact the upper parts of the Reuss and the Aar probably represent the rivers which brought down the great masses of Miocene gravel which now form the lowlands of Switzerland, and through which they subsequently cut the lower parts of their courses. These therefore must necessarily be of much less ancient origin; but even these valleys were as a rule excavated to their full depth before the Glacial period, and must therefore be of immense antiquity.

CHAPTER IX.

LAKES.

THE Alps are surrounded by a beautiful circle of lakes. We have on the north, besides many smaller ones, those of Constance, Walen and Zürich, Zug, Lucerne, Brienz and Thun, Geneva; on the south the Lago Maggiore, Lugano, Como, Iseo, and Garda, all seeming to radiate as it were from the great central mass of the St. Gotthard. I do not mention the Lakes of Neuchâtel or Morat, because they belong to a different category.

These great lakes are clearly not parts of a former inland sea. They stand at very different levels. The Lake of Brienz, for instance, is 190 metres above that of Geneva; that of Orta is 225 metres above the Lake of Garda.

But in considering the origin of these lakes we must have regard not merely to the surface level of the water, but also that of the bottom. When we give the level of a lake it is usual to quote that of the upper surface, but the bottom is perhaps even more important, and as we shall see from the following table, there is a great contrast between the two:—

	Surface Level.	Greatest Depth.	Bottom Level.
Constance	395 metres	252 metres	143 metres
Walen	423 ,,	151 ,,	272 ,,
Zürich	409 ,,	142 ,,	267 ,,
Zug	417 ,,	198 ,,	219 ,,
Lucerne	437 ,,	214 ,,	223 ,,
Sempach	507 ,,	87 ,,	420 ,,
Brienz	566 ,,	261 ,,	305 ,,
Thun	560 ,,	217 ,,	343 ,,
Geneva	375 ,,	309 ,,	66 ,,
Neuchâtel	432 ,,	153 ,,	279 ,,
Bienne	434 ,,	74 ,,	360 ,,
Orta	290 ,,	143 ,,	147 ,,
Maggiore	194 ,,	655 ,,	−461 ,,
Como	199 ,,	414 ,,	215 ,,
Lugano	266 ,,	288 ,,	− 22 ,,
Varese	239 ,,	29 ,,	210 ,,
Iseo	185 ,,	346 ,,	−161 ,,
Garda	65 ,,	346 ,,	−281 ,,

These depths are the more remarkable if we compare them with certain seas. For instance, the English Channel is nowhere more than 50 metres in depth, the North Sea, 60.

The original depth of the Lakes was, moreover, even greater, because the present bottom is in every case covered by alluvium of unknown, but no doubt considerable, thickness.

The Lakes of Neuchâtel and of Bienne only differ by 1 metre as regards the water level, but the Neuchâtel basin is 60 metres deeper than that of Bienne.

The great Italian lakes, as shown in the foregoing table, descend below, sometimes much below, the sea level.

The lakes, moreover, are in some cases true rock basins. In the case of Geneva, for instance, though the actual outlet is over superficial debris the solid rock appears in the river bed at Vernier only 10 metres below the surface of the lake, or 300 metres above the deepest part.

The materials brought down by the rivers have not only raised the bottoms of the lakes, but have diminished their area by filling them up in part, especially at the upper ends. It is evident that they were at one time much larger than they are now. The Lake of Geneva extended at least to Bex and perhaps to Brieg, that of Brienz to Meiringen, of Lucerne to Erstfeld, the Walensee to Chur, the Lake of Constance at least to Feldkirch, the Lago Maggiore to Bellinzona, that of Como to Chiavenna.

Moreover, the lakes of Brienz and Thun formed one sheet of water, as also did the Walensee and the Lake of Zürich.

Very slight changes might again greatly enlarge the lakes. For instance, if the narrow outlet of the Aar, somewhat below Brugg, were again closed, a great part of the central Swiss plain would be submerged.

The problem of the origin of lakes is by no means identical with that of rivers. We have not only to account for the general depth of the valley —this may be due to running water—but for the exceptional basin of the lake; running water produces valleys, it tends to fill up and drain lakes.

To what then are lake basins due?

It used to be supposed that many lakes were due to splits and fractures. I do not, however, know of any Swiss lake which can be so explained.

We may divide Lakes into four classes:—

1. Lakes due to changes of level.
2. Lakes of embankment.
3. Lakes of subsidence.
4. Crater lakes.

In many cases, however, a lake may be due partly to one of these causes and partly to another, and for convenience of description they may be dealt with under eight heads:—

1. Those due to irregular accumulations of drift; these are generally small and shallow.

2. Corrie lakes.

3. Those due to moraines.

4. Those due to rockfalls, landslips, river cones, glaciers, or lava currents damming up the course of a river.

5. Loop lakes.

6. Those due to subterranean removal of soluble rock, such as salt, or gypsum. These principally occur in Triassic areas.

7. Crater lakes.

8. The great lakes.

1. As regards the first class, we find here and there on the earth's surface districts sprinkled with innumerable shallow lakes of all sizes, down to mere pools. Such, for instance, occur in the district of Le Pays de Dombes between the Rhone and the Saône, that of La Sologne near Orleans, in parts of North America, in Finland, and elsewhere. Such lakes are, as a rule, quite shallow. They are due to the fact of these regions having been covered by sheets of ice which strewed the land with irregular masses of clay, gravel, and sand, on a stratum impervious to water, either of hard rock such as granite or gneiss, or of clay, so that the rain cannot percolate

through it, and where there is not sufficient inclination to throw it off.

2. Corrie lakes may be thus explained. Let us assume a slope (Fig. 64, *a, b, c, d*) on which snow and ice (*e*) accumulates.

The rocks and fragments falling from the heights would accummulate at *d*. Moreover, the ice would

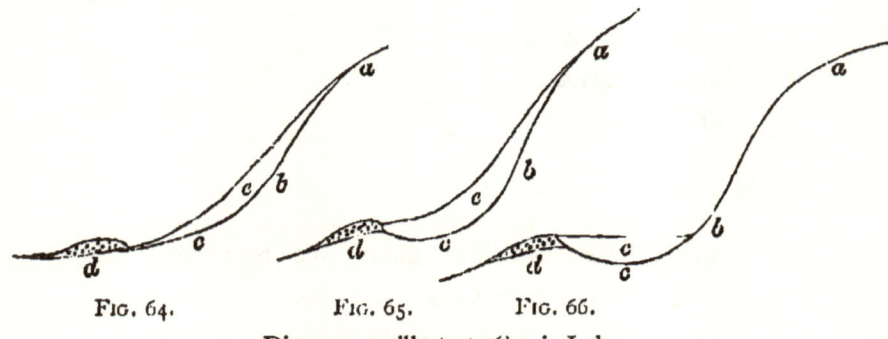

Fig. 64.   Fig. 65.   Fig. 66.
Diagrams to illustrate Corrie Lakes.

tend to form a hollow at *c* (Fig. 65) where the pressure would be greatest.

If subsequently the snow and ice melted, water would accumulate in the hollow (Fig. 66), and lakes thus formed are common in mountainous districts, where they have a special name—Corries in Scotland, Oules in the Pyrennees, Botn in Norway, Karwannen in the German Alps, etc.

3. A third class of lakes is that due to river val-

leys having been dammed up by the moraines of ancient glaciers.

To this cause are due the Lake of Zürich (in part), the Lake of Hallwyl, of Sempach, several of the Italian lakes (Iseo, Orta), and many others. In fact, most of the valleys descending from the Alps have, or have had, a lake where they open on to the Plain.

4. The fourth class of lakes were once even more numerous in Switzerland than at present. As cases of lakes due to rockfalls, I may mention the Törler See, near Zürich, and the Klön See in Glarus; among those due to river-cones the Sarnen See, and the lakes of the Upper Engadine; and as instances of lakes dammed back by glaciers the Lake of Tacul on the Mont Blanc range, and the Merjelen See, which is dammed back by the Aletsch glacier. In our own country the margins of such an ice-dammed lake form the celebrated "parallel roads of Glenroy."

5. Loop lakes occur along the course of many large rivers. The stream begins by winding in a loop which almost brings it back to the same point. The narrow neck is then cut through and the loop remains as a dead river channel, or "Mortlake." Again, when an island is formed in mid-channel, one

one of the side streams is often cut off, and forms a curved piece of standing water.

6. Subsidence lakes, as already mentioned, occur principally in Triassic areas. The gypsum or salt is dissolved away in places, and eventually the ground gives way, leaving funnel-shaped hollows.

Such a pool was actually formed near the village of Orcier in the Chablais in the year 1860. There had previously been a strong spring giving rise to a stream. Suddenly the ground fell in, forming a pond about 20 metres long and 8 wide. Three fine chestnut trees were engulfed, and the pool was so deep that at 20 metres no bottom was found, nor were even the tops of the trees touched.*

These hollows are generally small, though in some cases, as for instance the Königssee, the Lakes of Cadagno and Tremorgia in the Ticino, they are of considerable dimensions. Our Cheshire Meres are mainly due to the same cause.

7. Lakes occupying craters are far from infrequent in Volcanic regions, as for instance in the Auvergne, the celebrated Lake Avernus in the district of Naples, and the Maare of the Eifel. There are, however, no crater lakes in Switzerland.

* Favre, *Rech. Geol.*, vol. II.

8. As regards the greater Swiss lakes there has been much difference of opinion.

Ramsay and Tyndall maintained that they were rock basins excavated by glaciers.

Mortillet and Gastaldi* have suggested that the valleys were in pre-glacial times filled with alluvium, and that this soft material has been ploughed away by the glaciers.

"That glaciers rub down rocks," says Sir A. Geikie, "is demonstrated by the *roches moutonnées* which they leave behind them."

"Taking the case of a glacier," says Tyndall, "300 metres deep (and some of the older ones were probably three times this depth), and allowing 12.20 metres of ice to an atmosphere, we find that on every square yard of its bed such a glacier presses with a weight of 486,000 lbs. With a vertical pressure of this amount the glacier is urged down its valley by the pressure from behind."**

Indeed, it is obvious that a glacier many hundred, or in some cases several thousand, feet in thickness, must exercise great pressure on the bed over which it travels. We see this from the striae and grooves

---

* "Sur l'affouillement glaciaire," *Atti della Soc. Ital.* 1863.
** Tyndall, "Conformation of the Alps," *Phil. Mag.* Oct. 1869.

on the solid rocks, and the fine mud which is carried down by glacial streams. It is of quite a different character from river mud, being soft and impalpable, while river mud is comparatively coarse and gritty.

The diminution in the rapidity of motion of a glacier at the sides and near the bottom, which has been relied on as evidence that glaciers cannot excavate, shows on the contrary how great is the pressure.

The question has been sometimes discussed as if the point at issue were whether rivers or glaciers were the more effective as excavators. But this is not so.

Even those who consider that lakes are in many cases due to glaciers might yet admit that rivers have greater power of erosion. There is, however, an essential difference in the mode of action. Rivers tend to regularise their beds; they drain, but cannot form, lakes. As Playfair long ago pointed out,* a lake is but a temporary condition of a river. Owing in fact to rivers, lakes are mere temporary incidents. The tendency of running waters is to cut through any projection, so that finally its course assumes some

* Playfair's *Works*, vol. I.

such curve as that in Fig. 47, from the source to its entrance into the sea.

The existence of a hard ridge would not give rise to a Lake, it would delay the excavation of the valley; above it the slope would become very gentle, but no actual basin could be formed; we should have some such section as in Fig. 67. The action of a glacier is different; it picks out as it were the softer

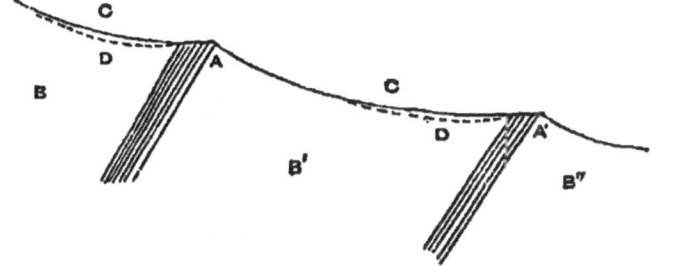

Fig. 67.—Diagram to illustrate the action of rivers and glaciers. A, A', Hard ridges; B, B', B'', Softer strata; C, C, Slope of running water; D, D, Slope of ice.

places, and under similar circumstances basins might be formed above the harder ridges as shown in the dotted lines, D, D.

In many of the Swiss valleys the pressure of the ice on its bed must have been very great. The Rhone glacier not only occupied the basin of the Lake of Geneva, but rose on the Jura to a height of 950 metres. The lake is 309 metres deep, so that the total thickness of ice must have been over

1000 metres. The greatest depth of the lake is opposite Lausanne, where the thickness of the ice would be at its maximum.

Moreover, the depth in proportion to its size is quite insignificant; Fig. 68 shows the height of the mountains, the thickness of the ice at the time of its greatest extension, while the dark line below gives the relative depth of the water, showing that after

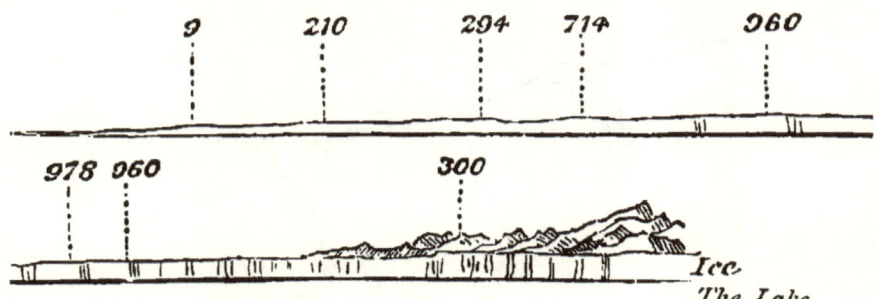

Fig. 68.— Diagram Section along the Lake of Geneva. The dark line shows the relative depth of the water indicated by the figures above.

all the Lake of Geneva is really but a film of water.

There are, however, strong reasons against regarding glaciers as the main agents in the formation of the great Swiss and Italian lakes. These have been pointed out with great force by Ball and Bonney, and Swiss geologists are not generally disposed to accept the action of glaciers as a sufficient explanation. They admit that glaciers grind and smoothe the rocks over which they pass, but deny that they effectively excavate.

The Lake of Geneva, 375 metres above the sea, is over 309 metres deep, and if we allow for the accumulation of sediment, its real bottom is probably below the sea level. The Italian Lakes are even more remarkable. The Lake of Como, 199 metres above the Sea, is 414 metres deep. Lago Maggiore, 194 metres above the Sea, is no less than 655 metres deep, so that the bottom is 461 metres below the sea level.

The difficulty thus arising, moreover, is not so much the absolute depth, as the absence of relative height above the Sea, so that there would be no sufficient fall to carry off the water.

Even if we suppose that the Sea came up to Lyons, still the distance from Lausanne being 180 km., the Lake must have been raised 300 metres to give even a minimum fall of 2 per cent.*

In the valley of the Rhone the upper level of the ice had a slight but regular slope. At Schneestock the upper limit was at a height of 3550 metres above the sea, at Leuk 2100, at Morcles near St. Maurice 1650 metres. But at Chasseron on the Jura the height is now 1410 metres, at Chasseral

* Forel, *Le Léman*.

1306, on the Salève 1330. This gives a slope of 2½ to 3 per cent only. Now in the present Swiss glaciers the slope is about 6 per cent. That of the glacier of the Aar, which is the least inclined, is 5 per cent. No doubt the greater the glacier the less is the inclination at which it can move. Still a slope of 3 per cent would seem quite inadequate. If, however, we suppose that the Alps had a relative greater altitude of say 1000 metres the difficulty would be removed, and the glacier would have a sufficient fall.

These and other considerations have led gradually to the opinion that while the valleys occupied by the Swiss lakes were mainly excavated by running water, the lakes themselves are due to changes of level which have raised parts of the valleys as compared with the river courses nearer the mountains.

Prof. Heim has suggested that the compression which elevated the Swiss mountains, and piled, as we have seen (*ante*, p. 91), more than double the original weight on this portion of the earth's surface, led to the formation of the great lakes. The mountain mass thus concentrated on a comparatively small area would from its enormous weight tend to sink somewhat into the softer magma below, which of course would have had in this respect the same effect as if the surrounding country had risen. The result would be to dam up the rivers and fill the valleys. For instance, in the Lake of Lucerne the bottom of the Bay of Uri is almost flat; it is evidently a river valley which has been filled with water.

In fact, speaking generally, the great Swiss lakes are drowned river valleys.

The relative subsidence of the mountains is no mere hypothesis.

There are, as we shall see, strong grounds for believing that the country round Geneva has been recently raised.

The old river terraces of the Reuss can still be traced in places along the valley near Zug. Now, these terraces must have originally sloped from the upper part downwards, that is to say, from Zug towards Mettmenstetten. But at present the slope is the other way, *i.e.* from Mettmenstetten towards Zug. From this and other evidence we conclude that in the direction from Lucerne towards Rappersdorf there has been an elevation of the land, which has dammed up the valley, thus turned parts of the Aa and the Reuss into lakes, and, as we shall see, considerably changed the course of the river.

Again, Professor Heim has pointed out that there

has been a comparatively recent elevation, even since the commencement of the Glacial period, along a line traversing the Lake of Zürich. This is shown by the fact, that while the lower terraces follow the general slope of the valley, the upper glacial deposits present for some distance a reverse inclination. M. Aeppli in his recent work* has described them in more detail. They are seen on both sides of the lake, between Horgen and Wädenschweil on the one side, and between Meilen and Stäfa on the other. They do not, however, exactly correspond on the two sides of the lake, because the zone of compression crosses the lake diagonally, commencing more to the south on the east side. For the same reason, while the compression has on the east side made the terraces slope towards the lake, on the west the slope is towards the hill. This curious fact was very difficult to account for, but is satisfactorily explained by the inversion of the terrace.

I had the great advantage of visiting the terraces on the west of the lake under the guidance of Professor Heim, and looking across we could clearly see those on the east side also.

Passing to other countries the case of the Dead Sea is very suggestive. From the lower end a long

* *Beitr. z. Geol. K. d. Schw.*, L. XXXIV.

depression leads southwards; it is evident that the Jordan once ran into the Gulf of Akaba and so to the Red Sea, and that a subsequent change of level has created the Dead Sea, which has a depth of 396 metres below the Ocean level.

The great American lakes are also probably due to differences of elevation. Round Lake Ontario, for instance, there is a raised beach which at the western end of the lake is 110 metres above the sea level, but rises towards the east and north, until near Fine it reaches an elevation of nearly 300 metres. As this terrace must have originally been horizontal, we have here a lake barrier, due to a difference of elevation, amounting to over 180 metres. But though the lakes may not have been excavated by glaciers it is probable that the process of filling up would have made much more progress had they not been for so long a period occupied by the ice.

The next question which arises is as to the age of the lakes. The valleys are now regarded by most Swiss geologists as pre-glacial, but the lakes themselves originated after the retreat of the glaciers.*

If these views are correct the larger lakes north of the Alps may be divided into three classes.

* Penck, *Vergletscherung der Deutschen Alpen.*

Firstly, the lakes of the Jura,—those of Neuchâtel, Bienne, and Morat, which occupy synclinal valleys.

Secondly, those of Hallwyl, Baldegger, Sempach, Greifen, etc., which are moraine lakes, the dams at the lower ends being moraines.

Thirdly, those of Constance, Zürich, Walen, Zug, Lucerne, Thun, Brienz, and Geneva, some of which are indeed partially dammed up by ancient moraines, but which are partly at least due to the lower ends of the valleys having risen relatively to the rest.

Dr. F. A. Forel has suggested * that this subsidence of the Central Alps also throws light upon the former extension of the glaciers. The present snow-line is at a height of say 2600 metres. If we assume the subsidence to have been 500 metres (which seems the minimum), and suppose that 900 metres have been since removed from the whole surface, certainly no exaggerated estimate, this would bring the snow down to the present line of 600 metres, which would involve a great extension of the Firn, and consequently of the glaciers. He considers that an elevation of 900 metres would bring the glaciers of the Rhone down again to the Lake of Geneva. The theory deserves careful study but is open to

* *Le Leman.*

the objection that the Glacial period is no mere local phenomenon, but seems to have affected the whole northern hemisphere.

In considering the great Italian lakes which descend below the sea level, one suggestion has been that they are the sites of the ends of the ancient glaciers, and their lower ends are certainly encircled by gigantic moraines. We must, however, remember that the valley of the Po is an area of subsidence and a continuation of the Adriatic, now partially filled up and converted into land by the materials brought down from the Alps. Under these circumstances we are tempted to ask whether the lower lakes at least may not be the remains of the ancient Sea which once occupied the whole plain. Moreover, just as the Seals of Lake Baikal in Siberia carry us back to the time when that great sheet of fresh water was in connection with the Arctic Ocean, so there is in the character of the Fauna of the Italian lakes, and especially the presence of a prawn in the Lake of Garda, some confirmation of such an idea.

However this may be, the lower ends of the lakes have been dammed up by glacial accumulations.

Further evidence, however, is necessary before

these interesting questions can be fully and definitely answered.

The Colour of the Swiss Lakes.

Switzerland owes much of its charm to the lakes, and the lakes owe their beauty in great measure to their exquisite colouring. In this respect they differ considerably: the Lake of Geneva is blue, but most of the Swiss lakes are more or less green, and some brownish. What is the reason of these differences?

The blueness is not due to, though it may be enhanced by, the reflection of the sky. Pure water is of an exquisite blue. Of all the Swiss lakes the Lake of Lucel in the Val d'Herins is perhaps the clearest, and it is of a lovely blue. Various suggestions have been made to account for the green colour of some lakes. The most probable explanation appears to be that suggested by Wettstein, and ably supported by Forel,* namely, that the blue is turned into green by minute quantities of organic matter in solution. Forel took water from several lakes, and thoroughly filtered them, but they retained their colour, showing that it was not due to particles in suspension. He then took a block of peat, and infused it

* *Le Léman*, vol. II.

in water, thus obtaining a yellow solution. By adding a small quantity of this to the blue water of the Lake of Geneva, he was able to obtain a green colour, exactly similar to that of the Lake of Lucerne.

He refers as a test case to the sister Lakes of Achensee and Tegernsee in the Tyrol. The basin of the Achensee is free from peat, in that of the Tegernsee peat mosses cover a large space. The former is a brilliant blue, the latter a lovely green. He concludes, therefore, with Wettstein, that the bluest lakes are those which are the purest; while green lakes contain also a minute quantity of vegetable matter, or peat, in solution.

This is, however, by no means the only cause to which water owes a green hue. Shallow water over yellowish sand is green by the reflection of the yellow light from the bottom. Again, after storms the water is often rendered thick and turbid. After the coarser mud has subsided the finer impalpable particles give the water a greenish hue, which, however, is only temporary, though it may last for some time. Finally, the water is sometimes coloured green in patches by microscopic algæ.

But though the blueness of lakes and seas is not owing to reflection from the blue sky, the brilliancy, beauty, and variety of tone and tints, the play of colour to ultramarine and violet, the constant changes and patterns varying with every breath of wind, in short the life and glory and beauty of the lakes are entirely due to the light of the sun.

The Beine or Blancfond.

If, on a fine, still day, we look down the Lake of Geneva from some neighbouring height, we see the azure blue of the deep water fringed by a clear grey or greenish margin. This is the "Beine" or "Blancfond" where the shallowness of the water renders visible the grey or yellowish tint of the bottom. Such a shallow fringe or margin encircles many of the Swiss lakes, and may be explained as follows: The waves gradually eat away the bank, giving rise to a small cliff and talus (Fig. 69 p. 256). The loose stones and sand are gradually rolled downwards, forming a slightly inclined terrace (Fig. 69, K, M) which finally ends in a steep slope. This terrace is known as the Beine or Blancfond. The depth of the Beine depends on that to which the water is agitated by the waves; it is less, therefore, in sheltered, and greater in exposed, situations. In the Lake of Geneva it ranges between 1 and 4 metres. It falls into two parts, the inner (K, C) due to erosion, and the outer (C, M) to deposition. The inclination of the outer slope depends on the nature of the materials; the finer they are the gentler it is.

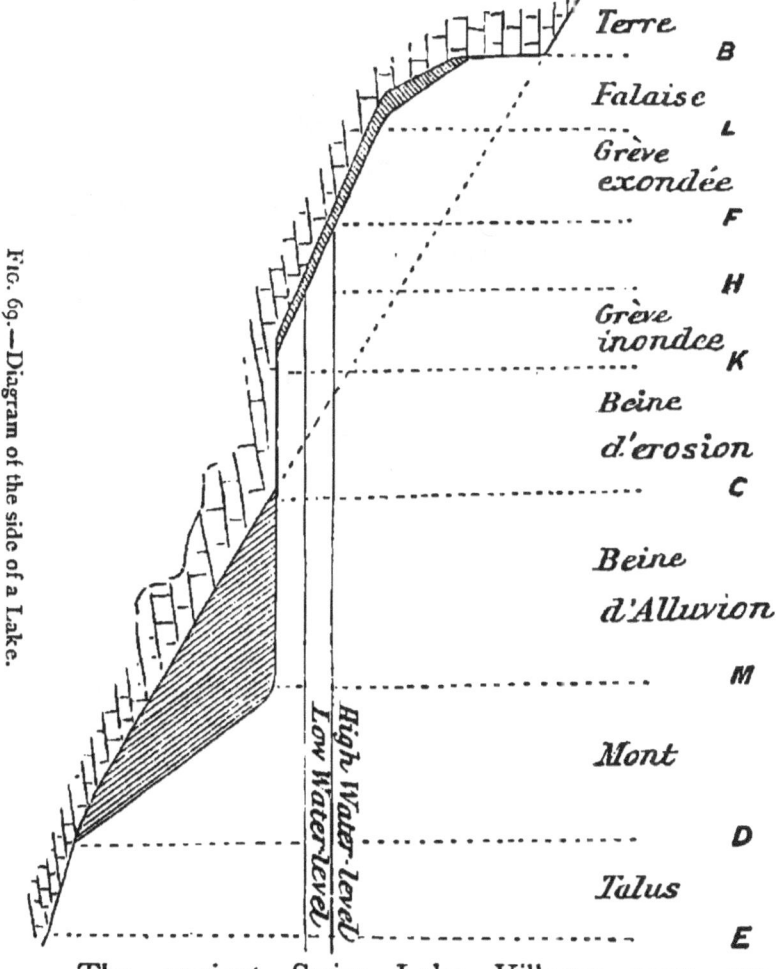

Fig. 69.—Diagram of the side of a Lake.

The ancient Swiss Lake Villages were constructed on the Beine, and this shows us how constant the level of the great Swiss lakes must have been for many centuries or even some thousands of years. Many of the lake villages belong to the Stone Age, and the stumps of the piles on which they were built still remain.

The platforms could not, of course, have been constructed over water more than at the outside 5 metres in depth, so that during this whole period the level of the lakes must have been practically what it is now. Indeed, the structure of the Beine itself shows that the level must have remained approximately the same for a very long period.

CHAPTER X.

ON THE INFLUENCE OF THE STRATA UPON SCENERY.

The character of Swiss scenery depends mainly on denudation and weathering, modified by the climate, the character, the chemical nature, the height, and the angle of inclination, of the rocks.

The total thickness of the sedimentary rocks has been estimated roughly at 200,000* feet, and as the whole of this was deposited in seas or lakes and was derived from former continents, we see how enormous the amount of denudation must have been, especially if we bear in mind that much of it has been washed down and deposited, then raised and afterwards washed down again; some of it moreover several times.

The principal forces which have disintegrated rocks are—(1) Water; (2) Changes of temperature; (3) Chemical actions; (4) Vegetation.

There are few rocks which are not more or less alterable by, or soluble in, water. It soaks in and

* Many of the beds, however, are not represented in Switzerland.

filters through innumerable crevices, dissolving some substances, especially when it is charged with carbonic acid, and leaving others. It also acts mechanically, for as it expands when freezing, it splits up even the toughest rocks, if only there are any crevices into which it can enter. In a dry climate, therefore, the slopes will generally be steeper than in a more rainy region. Even in the absence of water, changes of temperature have a considerable effect owing to the fractures which they produce by the successive contractions and expansions to which they give rise.

These, however, though the principal, are by no means the only factors in denudation. The roots of plants, for instance, have a considerable effect, insinuating themselves into the smallest crevices and, as they expand with growth, enlarging them by degrees. Yet, on the whole, the action of vegetation is conservative. It absorbs much of the rainfall, and the formation of torrents is thus greatly checked. Some of the French Alpine districts, and much of Northern Africa, have suffered terribly, and in fact been reduced almost to deserts, by the reckless destruction of forests.

Different kinds of rocks are very differently affected by atmospheric influences.

Siliceous rocks are liable to disintegration by

weather; but, on the other hand, the separate grains of sand or quartz are not only insoluble, but offer great resistance to mechanical action. Water, especially if charged with carbonic acid, can dissolve some Silica, but the quantity is insignificant.

Calcareous rocks are much more readily attacked. They often contain some alumina and siliceous nodules, which remain as a reddish clay with flints after the calcareous matter has been removed.

Argillaceous rocks cannot be dissolved, but they are in many cases readily reduced to fine particles and then easily removed. They generally contain some calcareous material, and when this is washed away, pores and hollows are left which let in moisture. Even when compressed into slates they often yield to the influence of moisture, and if sufficiently saturated sink into the form of mud.

Along the sides of valleys calcareous rocks often present steep, even vertical, faces (see *ante*, Fig. 44, p. 184, Valley of Bienne). Sandstones and Granite are generally less bold, and marly beds assume still more gentle slopes. The behaviour of argillaceous beds is more dependent on circumstances; if they are fairly dry they bear themselves well, but if they become wet they are very perishable.

So varied are the conditions that every mountain, even if the top only is visible, has a character and individuality of its own.

"Le profil de l'horizon," says Amiel, "affecte toutes formes: aiguilles, faîtes, créneaux, pyramides, obélisques, dents, crocs, pinces, cornes, coupoles; la dentelure s'infléchit, se redresse, se tord, s'aiguise de mille façons, mais dans le style angulaire des sierras. Les massifs inférieurs et secondaires présentent seuls des croupes arrondies des lignes fuyantes et courbes. Les Alpes ne sont qu'un soulèvement, elles sont un déchirement de la surface terrestre. Le granit mord le ciel et ne le caresse pas. Le Jura au contraire fait comme le gros dos sous le dôme bleu."

Not one of these varied forms is accidental. Every one of them has its cause and explanation, though we may not always know what it is.

The same configuration will of course look very different from different points of view. What seems like a sharp point is often the end of a ridge. The sedimentary rocks of the northern Alps (Rigi, Pilatus, Bauenstock, Sentis, Speer, etc.), often slope up gently to the summit, and then drop away suddenly in a steep cliff, frequently broken into a succession of steps which are rendered conspicuous by lines of snow. They give therefore what has been happily called by Leslie Stephen a desk-like form (Fig. 84), presenting broad, gently inclining plateaux, ending suddenly in a steep, almost perpendicular, precipice, which towers like a wall over the valley, such as the Diablerets, Wildstrubel, Gadmerfluh, Claridenstock, Tödi, Vorab, Balmhorn, Doldenhorn, Blümlisalp, etc.

In such districts still further denudation gives rise to ridges terminating in towers and teeth, sometimes of terrific wildness, as in the Engelhörner, or in the chain of the Gspaltenhorn. The calcareous Alps are also characterised by the numerous terraces, bands, pillars, and cornices. The precipices, as for instance on the Jungfrau and the great Wall of the Bernese Oberland, sometimes reach 2000 metres.

We might at first be disposed to anticipate that from their hardness and toughness the Crystalline rocks would be less liable to denudation than the calcareous. And in a sense this is true. In consequence however of these very qualities the drainage in Crystalline districts is mainly superficial, while in calcareous regions much of the rainfall sinks into the ground and is carried off by subterranean passages. In our own country we know that the chalk uplands, though cut into along the margins by deep combes, are seldom intersected by valleys, and almost all our railway lines leaving London have been compelled to tunnel through the Chalk. So also in Switzerland the calcareous strata form long continuous ridges, of which the great wall of the Bernese Oberland is a marvellous example.

Another reason for the extremely bold character of the calcareous mountains is that such strata are extremely stiff, and where argillaceous rocks would gradually bend, they break away and thus give precipitous cliffs.

It was at one time supposed that each kind of rock gave its own special mountain form. Such was the view, for instance, even of excellent observers, such as L. v. Buch and A. v. Humboldt.

It would, however, be quite a mistake to suppose that particular contours always indicate the same kind of rock. On the contrary, we find the same forms in different rocks, and different forms in the same description of rock. They depend greatly on the hardness of the rock, and on the angle at which it stands. Thus tower-like forms occur in Granite, Amphibolite, Sandstone, Conglomerates, Hochgebirgskalk, Dolomite, etc. The desk-like form which is so frequent in calcareous strata (see, for instance, Fig. 70 p. 266, on the right hand side) occurs also in some districts of Gneiss or of Nagelflue, as, for instance, at the Rigi (Fig. 84, vol. II. p. 55). On the other hand, the same rock may give a very different landscape. Thus Granite often assumes rounded outlines, but often also gives wild ridges of teeth and needles.

Gneiss summits with gently inclined beds are less steep and less pointed, while calcareous rocks if hard and steeply inclined assume not only wild but grand outlines. The Eiger is perhaps the finest type of a caleareous mountain.

On the other hand, in any given district similar geological structure will generally give similar scenery.

Steeply inclined strata as a rule produce bold outlines, while those which are more horizontal give a tamer scenery.

Still, where the rocks are very resistent, and denudation has been great, even horizontal strata may give very bold forms; of this we have a remarkable instance in the Matterhorn, a mountain left between two valleys, where the strata are but slightly inclined, and yet owing to their position and hardness give us the boldest and steepest mountain of the whole chain. In districts of the softer rocks we naturally miss the bold, steep precipices, the jagged ridges, and noble peaks, and must content ourselves with smiling landscapes and gentle undulations.

Another reason which affects the landscape in districts of sedimentary and Crystalline rocks is that the former crumble away more rapidly, and thus more quickly lose the rounded surfaces due to ice action. Thus, as we ascend the valley of the Reuss, where we leave the softer strata and enter the district of Gneiss, we also commence a scenery of knolls rounded by ice.

In calcareous districts "weather terraces" form a special feature (Figs. 44, 45 pp. 184, 185). They are due to a succession of rocks of different hardness and toughness, so that some strata weather back more quickly and take a gentler slope than others. Crystalline rocks are generally more homogeneous, weather more evenly, and consequently present more regular and continuous slopes. The Bristenstock, for instance, which towers over the Reuss, is a beautiful example. For a height of 2500 metres it presents an unbroken slope at an angle of 36°. Weather terraces are particularly conspicuous in certain lights, and especially in winter when there is snow on the gentler slopes. Even in summer, however, the contrast of vegetation is often striking, some lines being marked out by luxuriant grass or bushes, while others are comparatively bare. On Granite or Gneiss a good mountaineer can go almost anywhere, while in mountains of sedimentary strata he is stopped from time to time by an impassable precipice.

On the whole, when seen from a distance, the forms of the sedimentary mountains are more marked, more broken, and, so to say, more individualised.

The central Crystalline "massives" present very different forms. The desks, terraces, pinnacles, and cornices disappear, and we have noble pyramids. The ridges, moreover, are more jagged and serrated.

Fig. 70.—Ridge of the Gauli. Profile of the ridge from the Bächlistock to the Hühnerstock, showing the peaks of the granite rock and the desk-like slope of the calcareous strata forming the Hühnerstock.

Fig. 70 shows the contrast of a jagged Crystalline ridge and the desk-like form of the calcareous strata on the right (Hühnerstock).

In the splendid panorama seen from Bern the Crystalline mountain peaks (Finsteraarhorn and Schreckhorn, Breithorn, Tschingelhorn, etc.) can readily be distinguished from the calcareous mountains (Blümlisalp, Doldenhorn, Aletsch, etc.). The difference of character is also well seen as we ascend the valley of the Reuss from Fluellen to Andermatt.

On the whole, the calcareous chains of the Alps are wilder, the Crystalline grander.

Typical Gneiss often gives gentle rounded outlines. On the other hand, Sericitic Gneiss and Mica Schists, which often closely resemble Gneiss, show generally great readiness to fracture in sharp, knife-edge ridges, and very wild if perhaps less sublime forms. The Bernese Oberland owes both its great average height and the variety of its scenery to the combination of Gneiss with calcareous strata. The consequence is that it does not form an uniform range, like the Pyrennees, but a succession of individual mountains, presenting some of the noblest forms. In this district the Gneiss is inverted over the secondary strata, which it thus serves to protect. The result is that the weathering forms of both strata come into play, and thus produce endless variety.

Granite is regarded by poets as peculiarly resisting, and it is described as

> Stern, unyielding might,
> Enduring still through day and night
> Rude tempest shock and withering blight.

As a matter of fact, however, granites, as a rule, are very susceptible of disintegration. Granite mountains tend to gentle, rounded, and massive forms.

Rain, and especially water charged with carbonic

acid, acts on Granite profoundly. In many quarries where it looks solid enough it will be found to be disintegrated to a considerable depth, and even changed into a loose sand. This is due to the Felspar; the alkaline salts of Soda and Potash being decomposed by the carbonic acid, leaving the Silicate of Aluminium, the Mica, and the Quartz. It seems at first inconsistent with this that Granite ridges are often peculiarly jagged, but in such cases the Granite is steeply inclined, and the debris are removed as they form.

In other cases Granite shows a tendency to weather in convex, but somewhat flat shells, and to split vertically in two or often three different directions: it is divided, moreover, into horizontal layers at more or less regular intervals, thus forming rhomboidal blocks or pillars. Granite possessing this structure often assumes very bold, wild forms.

Protogine, though so similar to granite, generally gives a different scenery. It breaks up more readily into Aiguilles, and the divisional plains are more marked. The vertically constructed Protogine of the Mont Blanc range, for instance, has a different aspect from the chains which are composed of true granite.

The "Aiguilles" formed by Crystalline Schists,

as for instance in the Mont Blanc district, at first sight resemble dolomite peaks. The transverse lines, however, are not continuous, and the summits are even more pointed, though in many cases, as, for instance, the Aiguille de Charmoz, what seems a pointed needle is really a long, narrow crest. The materials are among the very hardest in existence.

Hornblende schist is sometimes quite pale, sometimes very dark. It often becomes reddish by decay of the ferrosilicate, so that many mountains of this rock are known as the Rothhorn, Rothfluh, etc. It forms bold, sharp ridges, and torn, wild, pointed peaks.

Porphyry, though rather rare, forms an extensive bed in the neighbourhood of Botzen, occupying an irregular strip, running from north to south, some 40 miles long by 12 wide, through which the outlet of the Adige has been cut. The great rounded walls of dull purplish-red rock, clothed in many places with brushwood, and supporting large upland plateaux of the richest herbage, produce a scene of singular luxuriance and beauty, especially when their tints are heightened by the glow of the setting sun. Beautiful as they are at all times, there is then something almost unearthly in their splendour; and no one who has not made an evening journey from

Meran to Botzen, or from the latter place by the gorge of the Kuntersweg, knows what treasures of colour the Alps can afford.

Dolomite is a magnesian limestone. The aspect of Dolomite mountains has been most aptly compared to ruined masonry, and it is often difficult to believe that the summits of dolomite peaks and ridges are not crowned by crumbling towers, castles, and walls built by man.

A square columnar formation is characteristic. The whole of the face shows transverse and vertical marking, the transverse lines running more or less continuously across the whole. "The jagged outlines of the crests form a principal feature for their recognition. The outline is usually 'embattled,' to borrow an expression from heraldry. The colours are marvellously beautiful—cream colour and grey predominate, but not to the exclusion of others. In the glow of sunset they are almost unearthly."*

The Upper Jurassic gives valleys a very characteristic aspect. It assumes a steep slope of from 40° to 60°. If the inclination is not above 45° it becomes covered with vegetable soil and often clothed with fir; but the steeper slopes are bare and arid, and are known as Châbles or ravières, giving an

* See Dent, *Mountaineering*, Badminton Library.

aspect of ruin and desolation, forming often a strong contrast with the brilliant vegetation below.

In calcareous districts the surface is sometimes quite bare and intersected by furrows attaining a depth of several, sometimes even as much as 30 feet. Such districts are known as "Lapiées" or "Karren". A good illustration is to be seen above the hotel at Axenstein on the Lake of Lucerne, where a portion of the rock has been uncovered. Another is at the Kurhaus on the Brunig. Rollier refers to a great erratic on the Lapié of Bonjean near Bienne, which has protected the rock below it, so that it rests on a flat surface in the middle of the Lapié. The Hohle Stein near Donanne is another case of the same kind. The Lapiées or Karren are extremely barren, but the rock generally contains some small percentage of clay, which is washed into the hollows and supports some scanty vegetation.

The Flysch gives gentle uniform slopes. The Nagelflue, in the familiar case of the Rigi, is an illustration of the desk-like form, with a steep escarpment towards the Bay of Küssnach and a gentle slope following the inclination of the beds from the Rigi Kulm to the Scheidegg. In other cases the Nagelflue gives a very complicated relief, sometimes forming mountain knots from which valleys radiate in all directions. Deep gorges, with perpendicular, almost overhanging, bellied walls, and abrupt terminations also frequently occur in Nagelflue districts, as for instance to the north of the Lake of Thun, on the Speer, and elsewhere.

Glaciated regions present us two totally distinct types of scenery: a central or upper of bare barren rock with rounded outlines (Fig. 32 p. 132), and a peripheral ring of debris in scattered heaps and long mounds.

These morainic deposits give a peculiar character to the scenery: the country is diversified and irregular, thrown into confused heaps and depressions, which, as the lower or ground moraine is very impervious, often contain small lakes. They occur especially in well-watered districts, and the rich network of rivers often take very devious courses. Desor has happily characterised such a district as "un paysage morainique."

The scenery is again affected very much in consequence of the influence of different strata on streams and springs. For instance, in a country of hard impervious rock we have numerous little runnels which gradually unite into larger and larger streams. On the contrary, in a calcareous district, especially if fissured, we find, as for instance in parts of the Jura and elsewhere, large districts with very few streams, and here and there copious springs, where the water is brought to the surface by some more impervious stratum. A glance at any geological map will show, for instance, that the districts occupied by the Upper Jurassic rocks are especially waterless, there being many square miles without even the smallest rivulet.

The distribution of springs naturally affects that of villages. Thus in several of the valleys of the Jura we find a row of hamlets along the outcrop of the impervious Purbeck strata.

The influence of different rocks upon vegetation is another way in which they affect the character of the scenery. The principal contrast is between Crystalline and calcareous strata.

Cargneule gives fertile pasturage, as do the Lower and Middle Jura owing to the quantity of Marl they contain.

The Cretaceous rocks furnish sweet but not abundant herbage, and the Lias is but moderately favourable to vegetation. The Urgonian districts are arid and barren, and can be distinguished even at a distance from the Neocomian, which bears a luxuriant vegetation.

Flysch supports a vegetation, vigorous indeed,

but of comparatively little value; the slopes generally bear dry grass and heather, while the flat ground is marshy.

High Alpine plants are often found on moraines, not so much from any peculiarity of the soil, as because of its coming from the heights.

Screes are generally bare from the continuous movement, which does not give plants time to grow.

Rockfalls.

Falling stones constitute one of the greatest dangers of the Alps. Tyndall was injured, and Gerlach killed by one. Many couloirs cannot be ascended without much risk, and the ancient passage up Mont Blanc, first discovered by Balmat, has been abandoned for another longer, but safer, route. Many of the steeper valley sides, as, for instances, those between Martigny and the Lake of Geneva, are furrowed by stone streams, which, like those of water, have their collecting-ground above their regular channel, and a cone of deposit below, which, however, stands at a steeper angle than that of a torrent. Many rock-faces have a continuous talus or scree of fallen stones at the base, which takes an angle of about 30°, and in some cases has almost climbed up to the summit. Along the valleys of the Niremont — Pleiades which abut on the Lake of Geneva at Montreux, the debris from the two sides meet in the middle, and attain a great thickness. One of the finest examples is that at the foot of the Diablerets, which rises from 2035 metres to about 2400 metres.*

The Glärnisch is nearly surrounded by rockfalls on its northern and eastern sides. They are mostly of interglacial age, and to one of them the Klönthalsee is due.

In the debris of rockfalls the edges of the stones remain fresh and angular, on many of them the surfaces show marks of blows, rubbing, hollows, and impressions, where they clashed against one another during the descent. They lie in wild confusion, large and small together, from fine dust up to rocks larger than a house. In some cases the originally loose materials have been subsequently cemented together into a breccia. The surface is very irregular, and often contains lakes, as, for instance, at Sierre in the Valais, and Flims on the Rhine.

The rockfall of Goldau from the Rossberg which

---

* Renevier, *Beitr. z. Geol. K. d. Schw.*, L. XVI.

occurred in 1806, and has been figured by Ruskin,* is well seen on the St. Gotthard line, between Lucerne and Brunnen.

Even more destructive was that of Piuro (Plurs) in the Val Bregaglia in 1618. After heavy rain a great part of the side of Mont Conto fell suddenly into the valley, and of 2000 inhabitants very few escaped.

At Flims ("Ad flumina," so called from the number of springs and streams) the road rises far above the Rhine and passes over an ancient rockfall, the greatest in all Switzerland, far surpassing that of Goldau. It blocked up the valley, thus forming a lake, and the Rhine has not even yet cut completely through it. The debris rise to a height of 700 metres on both sides of the river. They consist mainly of Malm interspersed however with blocks of Dogger, Verrucano, etc., and fell from the Flimserstein. The fall appears to have taken place between the first and last great extension of the glaciers. As in all rockfalls the surface is very uneven; and in the hollows are several beautiful lakes. The isolated eminences in the valley below Reichenau may be portions of another rockfall.

* *Modern Painters*, vol. IV.

Among other great rockfalls, or perhaps bosses of the solid rock which have been left by the river, may be mentioned those of Antrona Piana on the 26th June 1642, which destroyed the Parish Church and many houses, causing also much loss of life: those of the Diablerets in 1714 and 1749, of Montbiel in Prättigau in 1804, of the Dents du Midi in 1835, and that of Elm in 1881. The pretty little lake of Chède, on the road between Geneva and Chamouni, was filled up by a rockfall in the year 1837.

Nor must rockslips pass altogether unmentioned. Sometimes the movement is continuous, though very slow. In the chains of the Gumfluh, between Château D'Oex on the Sarine and the Diablerets, which are composed of hard calcareous rock on which vegetation establishes itself with difficulty, the cone of talus descends slowly towards the valley almost like a river.*

Again at Soglio in the Val Bregaglia a mass of detritus which has itself fallen from the steeper precipices, was for a long time, and probably is still, slowly moving downwards. The firs which grow on it do not stand upright, but cross one another at

---

* Favre and Schardt, *Beitr. z. Geol. K. d. Schw.*, L. XXII.

various angles, some being almost prostrate. The rocks below (Gneiss and Mica Schist) are inclined so that the edges retard the movement, which would otherwise be quicker and more dangerous.

Theobald tells us that in the summer of 1861, at the time of the melting of the snow, he was on a geological excursion near the Schwarzhorn in the Grisons when he gradually became aware of a strange roaring and crushing noise all round him. At first he paid little attention to it, but he at length found that the whole surface on which he stood was slipping downwards. He escaped as quickly as he could, but the movement continued, and about a quarter of an hour afterwards a great mass, 20 to 30 paces in length, precipitated itself over a precipice.*

Earth Pyramids.

Whenever we have a deposit of comparatively loose material with hard blocks, or layers, there is a tendency to form earth pyramids, owing to the looser material being here and there protected by a more or less tabular block of hard substance.

** *Beitr. z. Geol. K. d. Schw.*, L. II.

The most remarkable assemblage of such earth pillars is near Klobenstein, in the valley of the Katzenbach, near Botzen,* described and figured by Lyell; those above Viesch, and at Useigne in the Val d'Herins are other classical examples.

* *Prin. of Geol.*, vol. i.

END OF VOL. I.

www.ingramcontent.com/pod-product-compliance
Lightning Source LLC
Chambersburg PA
CBHW031928230426
43672CB00010B/1858